Y0-AAH-651

**DO NOT REMOVE
CARDS FROM POCKET**

**ALLEN COUNTY PUBLIC LIBRARY
FORT WAYNE, INDIANA 46802**

You may return this book to any agency, branch,
or bookmobile of the Allen County Public Library.

# EXPLORING
# HIGH-TECH
# CAREERS

By
*SCOTT SOUTHWORTH*

*The Rosen Publishing Group, Inc.*
NEW YORK

Published in 1984, 1986, 1988, 1993 by
The Rosen Publishing Group, Inc.
29 East 21st Street, New York, NY 10010

*Copyright 1984, 1986, 1988, 1993 by Scott Southworth*

**Revised Edition 1993**

*Library of Congress Cataloging in Publication Data*

Southworth, Scott.
  Exploring high tech careers.
  Index
  Bibliography: p. 185
1. High technology industries—United States—Vocational guidance.   I. Title
T21.S67 1984 001.64'023'73 84-4822
ISBN 0-8239-1502-6          ISBN 0-8239-1717-7 (pbk.)

*Manufactured in the United States of America*

# *Contents*

# CONTENTS

# Preface

This book is intended for those of you considering a high-tech career. Jobs are available that do not require an engineering degree, though employers are increasingly looking for a solid background with high-tech coursework.

This edition has been updated to reflect the changes in the volatile high-tech field. These updates include a new chapter, "Recent Changes in High-Tech."

My own experiences have contributed heavily to this book. I know what it is to hunt for a high-tech job without the credentials I would have desired (I have a degree in urban studies and planning and another in counseling). I understand the technical world, since I attended Massachusetts Institute of Technology. I know much about the job-hunting process. I have sought counseling on my own career decisions, and in turn I have counseled others on those matters.

Part I is an orientation to the whole field of high technology and high-tech jobs, as well as getting started in a high-tech job.

Part II is relevant to choosing a job category and describes major job categories in several areas.

Part III is relevant to job-hunting and getting a job. The final chapter is partly inspirational and partly common sense, stressing the value of a realistic, positive attitude.

The book is not meant to cover every detail of the job market in the high-tech industries, nor can it provide details on the job situation in specific parts of the country. Its suggestions and recommendations cannot be regarded as inflexible rules or golden answers that solve every problem. You must apply each suggestion with common sense and in light of your own situation and background. It is up to you to make your own success.

Even though the computer industry is more recession-resistant than some other industries, it too can suffer in times of severe recession, and at such times there may be job openings only for highly trained and experienced people.

I must acknowledge the valuable suggestions and advice of a fellow writer, Ray Tsuchiyama, and my other technical writing colleagues. Additional advice and help about the writing and publishing of this book has come from my mother (who has published her own work), Don Cox, and others. As always my wife, Zelda, has been a bedrock of understanding and support.

S. Southworth
Framingham, Massachusetts

# Exploring High-Tech Careers

# PART I

# 1

# *The High-Tech Age*

Technological developments affect almost every part of our lives. Within the past twenty years we have seen profound changes in technology, and the latest advances are employed in both work and play. Some of the changes are quite obvious. Office staff now create documents on word processors instead of typewriters; printed information often is sent by fax machine instead of the Postal Service; and computers are used instead of pen and paper to develop spreadsheets and design graphs. Other changes are less obvious. For example, the bottle of Heinz ketchup in your refrigerator used to be made by a crew of cooks in a kitchen full of bubbling kettles. Today that same ketchup is made by one cook using a computer screen and keyboard to control the robots and machines that make the ketchup. Whether obvious or not, technological developments have had—and continue to have—a profound influence on our lives. The high-tech age continues to reign, enabling us to do tedious or complex tasks accurately, quickly, and more efficiently.

Devices and technological approaches we once found comfortable are rapidly becoming outmoded as scientists and engineers develop new technologies. I can remember, as a student at Massachusetts Institute of Technology, proudly receiving my first slide rule. A slide rule looks a bit like a ruler. Marked with logorithmic scales, it is used to make rapid mathematical calculations. I remember learning to use it, learning how to keep it working smoothly, and using it in taking exams. I marveled at the cleverness of such a simple mechanical device. I thought it was something I would keep and use for many years. At that time (around 1968), it was hard to imagine the impact of computers.

Yet the first appearance of computers in the general marketplace was only a few years away. The slide rule was replaced by the hand-held electronic calculator, which in turn rapidly decreased in size while offering greatly increased computing power. I long ago abandoned my slide rule and now have a thin pocket-sized calculator with sixty functions.

Today calculators are commonplace. Computers, too, are appearing everywhere—in retail stores, in offices, and in our homes. Microprocessors are being used in products such as cars and clock radios. The personal computer is becoming less expensive, and soon every grade-school kid may be considered computer illiterate if he or she has not had some exposure to computers.

In some ways these developments are scary and confusing, requiring adjustments and new attitudes in our daily lives. They are also exciting and challenging. The high-tech age has entered the workplace. Many positions require a new job skill—ability to use the computer.

### Technical Stereotypes

Many people believe that workers in technical fields are introverted and quiet, antisocial and dull. They picture technical workers as nerds in wrinkled clothing, peering through thick-lensed glasses at complex configurations. The computer industry and the computer world are seen as dry, nonhumanistic, and narrowly technical. That is a stereotype, however, that does not apply in most cases. Usually, technical workers are bright and interesting. They are excited about the latest developments and eager to put them to use.

The technical viewpoint does tend to be precise, detailed, and focused on the problem at hand. That, in a way, is characteristic of any field of engineering. It is a kind of problem-solving approach that need not be carried over into one's personality or daily life. Also, many jobs with high-tech companies do not focus on such intensive engineering.

The narrow stereotype of technical workers applies less and less to people in the computer industry. Computers now are used by both big and small businesses, in academic institutions and government departments, and in the home. As more people use them, computers must become less technical and require less specialized knowledge to use.

### Computer and Programming Developments

Twenty years ago computers were not easily accessible to high school and college students. They were very expensive, so few schools had them. Students lucky enough to have computers at school had to take special courses in how to use them. The programmer had to adapt to the computer and to its demands and needs, putting his or her own needs second.

At that time, programs had to be submitted for batch processing. It was tedious to develop and debug programs, since you had to wait days to get your results. And often all you got back was your original deck of programming cards and a bunch of error messages. Programming had to be very precise, and minor mistakes could kill an entire run. Programmers had to look through long printouts of their coding, trying to track down some little error. They may have typed an "0" (the number zero) instead of an upper-case "O" (the letter), or vice versa, the only problem being a slight difference in shape.

Today computers are more accommodating. Programs can be written in an interactive mode, meaning that the computer interacts with the user, making statements or asking questions to which the user responds. Computers are more "friendly." You can write a program on a terminal, run it in a time-sharing mode, and get your results back quickly. You can repeat the process to refine and debug the program. Error messages tend to be more helpful, less jargonistic, and more easily understood. The computers are more forgiving of small errors such as inserting an extra blank space in a sentence. Some computers even point out errors line by line as you type in your programming instructions.

### The Availability of Computers

Almost everything in the future is expected to go digital. The binary language of 1's and 0's used in computers will be used in hundreds of other items. Microprocessors will soon be put into any appliance that has a conceivable use for them. The dishwasher, the clothes washer, the blender, the toaster oven, all can use the intelligence provided by the computer. Already, clock radios can be programmed to repeat the alarm at certain intervals or to wake the husband with one music station and then the wife with another. Microwave ovens can be programmed to thaw, bake, or warm various foods. Coffeemakers can be programmed to have steaming, fresh coffee available when you wake. Slowly, computers are being used everywhere, making our lives easier, and giving us more free time.

Speaking computers now are used by the blind and physically disabled. People unable to use a keyboard or see a computer screen can carry on simple conversations with the computer. They can give it orders by talking to it, and it can respond, telling the user what is happening on the screen. Computers already can turn lights on and off and control the thermostat. Soon they may be able to start dinner.

Computers are becoming readily available. Access to data banks, news reports, and the power of large computers is available to home computers by use of a modem. A modem is a device that allows a terminal to be connected to a computer over phone lines. People using a modem in their home can call up articles from the *New York Times* or *Wall Street Journal* on the computer screen. Although expensive now, this service should decrease in price over the years.

The basic cost of computer hardware already is decreasing, mostly because of development of the integrated circuit, which is contained in a chip. As technology advances, circuitry in the chip can be made smaller and smaller, and more circuitry (and computing power) can be placed in each chip. Even when more power is packed into a chip, the cost of manufacture does not increase proportionately. As a particular chip becomes popular, it can be mass-produced so that the cost becomes quite small. Simple microprocessors (the chip only, without accessories) can cost less than fifteen dollars.

Children today are growing up in a dramatically different environment. Many grade schools use computers to teach reading and math. Students learn about computers and computing at a very young age. Calculating power already available would astonish the slide rule–using engineers of a previous age. Terminals are becoming familiar objects, available in all kinds of job settings, such as supermarkets and department stores. Soon counter sales personnel will be able to make instant credit checks or inventory checks. Stores and companies that cannot do such things will seem backward and clumsy in their operations.

These advances will not all happen immediately. Sometimes developments are slowly put into use. Even with the growing use of computers there seems to be some resistance, particularly by people who are not used to using them.

### High-Tech Jobs

Yet the good jobs of the future will be available to those who are willing to try computers and become familiar with them. The jobs will be available in the high-tech industries themselves, as well as throughout society. Already, most businesses use computer power to improve efficiency. Companies use fax machines, word processors, cellular phones, mainframes, and minicomputers to perform daily tasks. People who can understand computer instructions and master the commands they need to know are becoming valuable and well paid.

Many of the people using computer technology today do not have highly technical educations. Many have high school diplomas or college degrees in nontechnical fields like English or history. These people have taken advantage of the greater accessibility of computers to make themselves valuable workers. They find themselves suddenly becoming "technical" people (though they do not really feel any different). They are proud of their skills, and with their technical knowledge they often vastly improve their job status, wages, and opportunities.

Many people currently employed in technical jobs do not have degrees in computer science or engineering. But they have taken appropriate high-tech courses or training programs. High-tech employers are more and more looking for a background of relevant high-tech courses. A high-tech degree is becoming more important, although any technical experience can be a plus.

What is happening in the computer industry is not so different from what has occurred throughout the twentieth century with each technological advance. The bicycle mechanic looked at the odd contraption called an automobile. He began to repair autos, learned as he went along, and gradually found himself in a thriving new business. With each new technological product, vast numbers of people who were willing to learn entered new fields. They helped to build whole new industries.

The computer industry still is relatively young. It is being built both by the technical experts and by those who are willing to learn. The industry is growing fast. Making the computer accessible to the average person opens vast new markets and requires armies of people to design and build the computers, to operate them, to write new software, to sell them, and to train people to use them.

If personnel interviewers in the computer industry waited to fill jobs with people having the ideal training and background, they would find few candidates. Instead, they concentrate on hiring good people who have some technical background but not quite what was wanted. The computer industry needs far more technical people than are available. The industry, in fact, is built upon people who did not have the "right" technical background.

Many more workers are needed than just highly trained engineers or programmers. Computer industries need people in sales, training, personnel, public relations, technical writing, teaching, field service, illustration, and other less technical areas. These positions do not require knowledge of how to design an arithmetic logic unit or write a relocatable assembler. They need some technical understanding, but not five years of schooling as an engineer.

The computer industry needs people from a broad background. Computer companies cannot operate from the old narrow, highly technical viewpoint, when it was assumed that any company that bought a computer would have to train or hire highly technical people to run it. The industry is enriched by people who have varied backgrounds and come from other kinds of educational and job experience.

The computer industry needs that kind of people, people who can enrich the design of computers to make them more attractive and easier to use, people who can find new artistic uses for computers, people who relate easily to the average person and understand what that person needs and wants in a computer. Nontechnical people have a great role to play in ensuring a more humanized computer industry. Their perspectives can make a great difference in what happens in the future.

# 2

## Advantages of a High-Tech Job

A high-tech job can be both rewarding and exciting. The high-tech companies are continually introducing new products. Even if not working on a high-tech research project directly, you may still find yourself with an "in" on the latest developments in computers. You may find out about new developments before the general public does and have a better understanding of what they mean.

These companies are growing and expanding. They continually need new personnel and compete for the people who are available. Salaries and wages can only be driven up by such a situation. The rewards in the form of direct wages, benefits, and opportunities for advancement are great. People moving into the high-tech field are often surprised by the generous salaries that are considered the norm. New and entry-level workers often make the mistake of not being aggressive enough about the salary that they request.

The high-tech industry offers good opportunities for job advancement. Working conditions are good, and arrangements such as flexible working hours or working at home are often available. Professional contract workers can work as much or as little as they like and take leisurely vacations while still commanding an excellent hourly wage. Although some of the people in the industry are narrow, technical types, there are also great numbers of interesting, intelligent people with broad interests.

One of the main things I have noticed about working in the high-tech area is a good outlook on life. Even in the midst of a recession, people generally have a positive and hopeful attitude, strengthened by a good financial position and job security. Of course, no industry can provide a guarantee of no layoffs and unending growth, but the computer industry has done well in those areas.

### The Strength of the Computer Industry

The point is that the high-tech industry is healthy and strong. The new products and the great expansion of computers into our daily lives means greater and greater demand, not only for small personal computers and terminals but also for the large computers. Very often large computers provide the computing power for individual terminals by direct connection or over phone lines. The greater availability of computing power means that companies rely on it more and more; having computing power available becomes a new standard of doing business.

Computers can save businesses much time and money by handling repetitive and tedious clerical tasks. Labor, even clerical labor, is expensive for any company. Computers can do many of the tasks performed by clerical workers faster and more accurately. Small-business owners can see other benefits, such as obtaining information that was too difficult to obtain before. They can use the computer to generate weekly reports about how their company is doing. The computer also can provide accounting and financial analysis.

Anyone who has done a weekly payroll for a small

company realizes what a difficult job it is when done by hand, especially under a deadline. The computer can handle such a function quickly and efficiently.

The advance of computers tends to put many clerical personnel out of work. On the other hand, new jobs are created as people are needed to operate and program the computer and maintain it (without even considering designing and manufacturing it).

Business executives often prefer to invest in computers rather than maintain a large staff. The purchase of computers has both tax and budgeting advantages. Managers usually are under pressure to keep labor costs down, and computers can help them reach this goal.

Computers are in high demand because they can reduce business expenses. A company's capital investment in computers can therefore be paid back quickly. Business also has the money to spend (if the investment looks good) more often than research institutes and universities, which were once the main users of computers.

Even if a small company has a computer, faster, more efficient, and cheaper computers are always being built. The executive with a computer must always consider whether to invest in a new model, partly for greater efficiency and partly because many of the company's competitors are obtaining greater and greater computing power.

Computers are becoming more efficient. The designs are being improved, and new manufacturing processes allow more circuitry to be packed into the same area on a chip. The process is expected to continue for at least the next ten years. Thus we can see the underlying strength of the industry.

Although demand for computers has diminished in the corporate area since the 1980s, demand is higher elsewhere. Smaller businesses are buying their first computers, while the corporations are focusing on maintaining and advancing their computer power. Thus, the computer industry has continued to grow.

This continuing growth translates into demand for well-trained personnel. The growth is not only in manufacturing but also in administration, user support services, design, training, field service, and systems analysis. This ensures a steady supply of job openings, both at the entry level and higher. The demand for good personnel keeps wages and benefits high and ensures the worker of opportunities for both salary and career advancement.

### Salaries, Wages, and Benefits

Salaries and wages are higher in the computer industries than in most others. Entry-level jobs for programmers and technical writers are often in the range of $20,000 to $30,000. Beginning engineers can earn as much as $30,000 to $35,000.

Wages for hourly paid workers such as assemblers can still be around $15,000 to $25,000 or higher for experienced workers or with overtime work.

Experienced engineers and programmers or managers can earn much higher salaries—approaching $40,000 to $80,000 a year.

These wages are often shocking to people working in such areas as teaching, counseling, social work, and government. Even university professors who hold PhDs are disturbed to find that people with bachelor's degrees in the high-tech industries are earning as much as they do.

Benefits in the high-tech field can be generous. They can include such things as health, dental, disability, and life insurance, all partially or fully paid by the company. Stock options are often available. Vacation policies can vary widely, but in most companies employees receive at least two weeks per year.

Companies keep an eye on each other to see what is being offered in benefits, which can add up to a substantial amount of money. A basic figure often used is 30 percent of base salary.

If you want more vacation time, you might consider working as a contract employee. The contract worker works for a specified period of time (such as three months) and has the opportunity to take lengthy vacations between periods of work. He or she usually receives a higher hourly wage but no benefits. Certain people like this flexibility and time off and do not care about the benefits.

### Recession and Inflation

High-tech companies are not completely resistant to the effects of recession. They are sometimes forced to lay people off, but more often the effect is a decrease in growth rather than a company's actually losing money.

Some companies are heavily tied into government contracts, and layoffs occur when a contract runs out and a new one has not yet been obtained. Paradoxical situations can occur, as when one division of such a company lays off personnel while another division has just obtained a new contract and is hiring personnel.

Many companies can avoid large layoffs, with the effect being more of a slowdown, especially in new or ambitious projects. One company caused a hullabaloo by shutting down a manufacturing plant for a week during the Christmas holidays. Employees were requested to take their vacations at that time; otherwise they would not have been paid for that week. It is interesting that in other industries this has been common practice for years.

In other high-tech companies, ambitious projects may be cut back in scope or abandoned altogether. In such case, personnel hired specifically for the project may be

**15**

laid off, but the number is quite small when compared to the large layoffs of, for instance, the automobile companies.

In most cases the average worker feels the recession in less severe ways. In companies that pay tuition for courses that employees want to take, the employee may have to offer better justification for such a course or delay taking it. Pay raises may be delayed or less generous. The worker may have more work to do, because the company is limiting new hiring.

Inflation is another economic evil that affects high-tech companies as well as other companies. In inflationary periods, demand for products is usually high, and such periods are boom times for the computer industries. Demands for personnel are especially high, and that tends to keep salaries high.

While high-tech companies do not necessarily offer cost-of-living increases, raises are given for merit and performance. In inflationary periods, raises tend to be higher. There is no guarantee that a person's salary will rise faster than inflation, but in many cases it does keep up with or even exceed inflation. Of course, in some unionized situations in which the worker receives cost-of-living increments, the wages do keep up with inflation.

### Advancing in a Computer Career

The opportunities in the high-tech industries can be many. The best goal is to establish yourself in a job and gain one or two years' experience. Then other opportunities within the company or within the industry will open up. Departments that are expanding rapidly need supervisors and managers, who often come from within the department.

Some companies offer large-scale programs in sales and programming to train personnel for better positions.

Again, once you have established basic credentials of work in the computer industry, you can apply for higher-level jobs in your company or for jobs in other parts of the country. Most metropolitan areas have some high-tech industries, and so do some rural areas such as southern New Hampshire.

### Working Conditions

The working environment is usually good. Companies often have well-equipped offices in modern buildings in suburban or rural areas away from the congestion of downtown areas. This is especially true in Massachusetts, where the high-tech belt surrounding Boston has been built up in small towns or the wooded and agricultural countryside. The hectic, congested commute along crowded expressways is thus avoided. Many Boston-area high-tech commuters use lightly traveled roads through a pleasant New England countryside. (To be honest, however, some have to travel the now heavily congested Route 128.)

Most professional workers have their own office, often with a computer or word processing terminal in the office itself. In other cases terminals are available in a common area nearby. The office is usually one with shoulder-high partitions that afford some privacy.

Working hours are often flexible, but that should be checked with each company. I remember interviewing for an entry-level job as a computer programmer. The company required that the worker be there from 10:00 to 3:00 Tuesday through Thursday, during which time meetings could be scheduled. He or she had to set up a regular weekly schedule each month indicating when he would be there to make up the rest of his forty hours.

Other companies offer various types of flexible plans based on the usual eight-hour routine (plus one hour for lunch) but allowing the employee to come in one or two hours later or earlier as desired.

Sometimes employees can work on a terminal based at home, coming in one or two days a week to attend meetings and keep in touch with their department. I know one programmer who worked in western Massachusetts for a company based in eastern Massachusetts. A software technical writer was able to work in his downtown Boston apartment for a company situated in the Boston suburbs.

Such arrangements are not automatically available and can vary widely. If they are important to you, you should check them out on a company-by-company basis. I also emphasize that in an entry-level job you may have no choice but a typical work schedule. Later you may be able to search for a job that offers the schedule flexibility or home work arrangements that you may desire.

One thing that should be mentioned is that many of these companies are set up on an 8-to-5 schedule—what I would call an engineering schedule. In the 9-to-5 schedule the lunch hour is included in the 8-hour day; in the 8-to-5 schedule the lunch hour is in addition to the 8-hour day.

Manufacturing and hourly paid personnel usually do not get flexible scheduling plans; they do, however, get the advantage of overtime. Workers such as assemblers often have a pleasanter work environment than that of many heavy industrial workers. The jobs often are not tied to a moving assembly line, but are sit-down jobs that require some intelligence and thought while working.

One example is work on line-calibrating equipment. Each piece of equipment has to be dealt with individually, and the worker is not under restrictive deadlines for each piece.

Another job requires the worker to construct a simple device based on circuit diagrams and instructions provided. The devices require some thought and intelligence in the making. That is not to romanticize jobs in the high-tech industry. There are still many jobs that are boring and repetitious, but there also are many that are not.

### People in High-Tech

Many of the people in the computer industries are interesting and intelligent and have broad backgrounds. Among people I have known, one repairs violins and is learning to make them, another is a big fan of square dancing, another plays in a rock group in her spare time, another is interested in sailing and has his own sailboat, another is interested in starting his own personal growth and visualization group, and another does photography.

I have also been surprised at the amount of artistic effort I have seen—work done by high-tech workers, not professional artists. I remember seeing a fabric in an intricate abstract pattern that was colorful and interesting. Later I realized that it was based on the type of pattern that appears on printed-wiring boards. Another example was the Boston skyline etched on a copper plate, which had been created by use of the photographic and etching facilities of a printed-wiring board manufacturing plant.

Various kinds of computer graphics are becoming more and more available. Many workers, especially designers, work with powerful graphics systems. These systems can produce dramatic visual and artistic effects and can be a lot of fun to play with.

Many high-tech workers—especially computer programmers, it seems—have a great interest in science fiction and fantasy. Offices are sometimes colorfully

**19**

decorated with posters from "Star Wars" or "Star Trek." Programmers sometimes name their programs after wizards or dragons, implying the magical or fantastical element in computers. This can be experienced after long hours of work when a program finally works correctly and seems to magically produce an impressive result with effortless ease.

# *3*

# *The Entry-Level Job*

The entry-level job can be one of the most difficult to find. Often these jobs are less interesting or carry less exciting or critical assignments than more senior positions.

Yet the entry-level job is the ticket to success, for once you have it, other opportunities begin to open up. In the entry-level job you are establishing a background and job experience that are invaluable in applying for or progressing to higher-level jobs.

This experience is the all-important point that personnel people look for. Almost anyone can learn what is needed to handle a technical job, but what the personnel interviewers look for is definite proof, and no proof is better than on-the-job experience. They also want to avoid putting their company in the position of doing on-the-job training (unless the job is specifically set up for that purpose.)

When looking for an entry-level job, you do not have that on-the-job experience. That is where a good résumé, good background, and doing well in an interview are important. Part 3 of this book deals with those things. Setting up a job strategy is important, and personal contacts can be invaluable.

## Entry-Level Salaries

Money is a tricky thing with an entry-level job. Sometimes the pay is fairly high, close to the level for a person with one or two years' experience; other times it is quite low. People applying for these jobs are sometimes put off because the wages are not what they expected.

Still you may find the pay quite adequate. I know a woman who was assistant director of a museum with several years' experience in her field. She got fed up with her salary situation and took a job with a major computer company; her entry-level salary was about the same as her museum salary.

Companies sometimes pay low salaries to incoming employees on the grounds that it will cost money to train them or that they will not be very useful or valuable to the company during the first year. Whether this is true varies greatly from job to job. You can be useful from the first day of work. Industriousness and willingness to learn can far outweigh many of the disadvantages of being inexperienced.

The problem is one that continually confronts the high-tech industry. Very often it is not possible to hire someone who has the level of technical background that may be desired. Rapidly expanding departments in a company often need to bring in entry-level personnel, and whether or not management so designates it, a lot of on-the-job training is done.

In some cases, a company may want to hire you because they can pay a lower salary. They think they will get enough work out of you to justify your pay even if you are still learning the job. A more experienced worker might get more work done, but it is not always clear that the additional work will correspond to the additional salary. A $40,000-a-year employee does not always do twice as much work as a $20,000-a-year one.

But then there are always situations in which the nature of the job demands the more experienced person.

### Utilize Your Background and Experience

When you are applying for or considering an entry-level job, you should relate your past experience to the job you are considering. Personnel departments sometimes look for more formal credentials such as a college degree or on-the-job experience. Sometimes they are amenable to looking at various other kinds of experience.

For instance, I once worked repairing copying equipment (for a chain of copy centers). Most of the work was mechanical, involving nuts and bolts and replacing broken parts, but at times I had to get into the electronics and track down which relay or switch was causing a problem. This required that I look through the circuit schematics to see what relays and switches might affect a particular area of the machine. I never thought too much of it, since I had a simplified schematic. I had also become familiar with the way the machine operated and had been able to consult with experienced repairmen so that the schematics soon became comprehensible to me.

Later when I was working at a high-tech company I was talking with an engineer. He seemed impressed with what I had done in the previous job and remarked, "Well, you must have been able to read schematics."

That was a simple enough statement, but I realized that I had in fact done something that should have been on my résumé. Schematics are, of course, the basic plans for any piece of electronics, and before a new computer is built the schematics must be worked out in detail. Electrical engineers must be able to read as well as write such schematics.

I should have included some reference to this experience on my résumé. Presented in the right way (even without exaggeration), it would have been a worthwhile skill. For certain jobs, such as computer repair (field service), I should definitely have included it.

In Chapter 15 I discuss getting the most out of your background. Briefly I can mention here several kinds of valuable experience: any that includes auto repair, which many people do on weekends for their own car; any kind of experience with a computer terminal or keyboard, including working with a home computer and any kind of word processing. Office workers often have some experience with a computer on the job, even if it is only giving the computer a few instructions to produce a preformatted accounting report.

These types of experience are all valuable, whether they are used in a résumé or in an interview situation. You must consider what will look well on a résumé for a particular job you are interested in, and what an interviewer seems to respond to.

### Use Your Present Job as an Entry-Level Job

You can use a current job as an entry-level job. That is, you can find a way to gain computer experience in your current situation, thus making it an entry-level job.

For example, many offices have small computers or terminals connected to large computers. Even if you don't have access to the computer as part of your normal routine, you may be able to talk the supervisor or boss into giving you access. You may be able to think up a project, for instance, some kind of program or analysis that will produce a useful report.

Usually manuals are available describing how the system works and how to program it, or there is an experienced person in the office who can help. If a project is useful, and especially if you are willing to spend some of your own time on it, the higher-ups often will look favorably on it.

These possibilities vary greatly from one office to another. Some bosses are willing to give you official support, whereas others are very negative. If this boss is negative, there is probably not much to be done about it; but if there seems to be no opportunity to learn and advance in your job, you should consider leaving it for a more open workplace (whether high-tech or not).

I would not encourage you to try to operate the computer without official support. Though not very likely, it is possible that company files and data could be damaged or lost, or other problems could occur. It is always important to find out what not to do and to avoid certain critical commands (such as deleting files) until you understand the command thoroughly.

A big problem here can be stereotypes and bias against women or minorities. Managers often see secretaries or clerical workers (who are often women) as being of low ability and have difficulty seeing them in higher positions, such as programming. There is no good answer for that except to persevere or to look for a more open job situation.

The key here is that computer expertise is in high demand and so is expensive. If you can show some ability and help a computer to run better, management may suddenly become very supportive of letting you learn more. Nothing succeeds like good solid results.

One advantage of an entry-level job is that you gain access to company training programs or tuition-paid outside programs. The job may not be what you really want, but it is a beginning. Many companies support training for their employees and see it as one important way of meeting their needs for technical workers, by upgrading the skills and knowledge of workers in lower-level jobs.

# 4

## Advancing from the Entry-Level Job

The key to getting ahead is to take advantage of your present situation. Thus, if you are in an entry-level job in a computer company, you are already in a good position to advance (within a few years) to a higher position. You need to take advantage of company-sponsored training or company-paid coursework, and you need to be ambitious and energetic when you apply for a new job.

### Examples of Advancement

I have known many people who have worked their way up within a company and advanced to better jobs. Each case was different, and there is no magic formula for success. Here I summarize some of the cases:

- Line technician to engineer. This friend of mine worked calibrating equipment as it came off the assembly line. He had a scientific degree but not in a field closely related to electronics. He started working on engineering projects, applied for a program to obtain a master's degree in electrical engineering, and was accepted. The company

provided support by paying his tuition and living expenses for one year. He is now working as an engineer, is receiving greater and greater responsibility for individual projects, and will soon complete his thesis and receive the master's degree.

- Field-service technician to manager. This woman began working as a field-service technician repairing computers for a large mini computer firm. Within a few years she was able to advance to a supervisory position and then to manager. There is great demand for competent field-service people because of the number of computers in use. As more and more small personal computers are built, there will be even greater demand in this area.

- Typist to instructor in word processing. This veteran was able to use skills that he learned in the military service to move into a better job. In the Army he was required to type up radio transmissions while they were being transmitted and thus became a fast and accurate typist. After the war he began as a typist for a computer manufacturer, then moved into word processing. After several years he was promoted to word processing instructor. His responsibilities included traveling throughout the country giving courses and developing new course materials.

- Programmer to manager. This employee worked as an assembler-level programmer on computer operating systems for a large computer company. His work was considered valuable, and he was promoted to manager of a small department. Eventually he tired of working in a large company and found a new job analyzing operations system coding for a small computer company.

- Teacher to instructor in computer programming. This woman who taught high school mathematics was able to get a job as a computer programmer. Because of her teaching background, she was soon able to become an instructor in the programming field.
- Layout designer to instructor. This person started his career laying out printed-circuit board designs based on the schematics given him by an engineer. He was able to move into a job instructing others in printed-circuit board layout and also became involved in writing instruction and reference manuals.
- Housewife to technical writer. A single mother returned to school while continuing to raise her two children. When she had earned her college degree, she was able to get work as a technical writer and is now pursuing a career in that field.

In none of these cases was anything given on a silver platter. Each person had to work hard, and most changed departments within a company or changed companies altogether to get ahead. Each person built on the experience he or she had to gain a job, and then built on additional experience to gain a better job.

### Changing Jobs Within a Company

You need to be able to change jobs. You may tire of a job, and to grow and develop you may need a change of responsibilities and activities. Changing jobs within a company can be another way to get ahead. The personnel department is familiar with your record, and even if it is not perfect, you are still a "known quantity" and your level of skills and ability is well understood.

Companies have different policies, but many give preference to employees. Often open jobs are posted on

a bulletin board in a particular plant or office building. Jobs at other locations may be harder to find out about. You may want to check a company's policy on these matters when you apply for an entry-level job.

One of the main advantages of staying with a company is that you keep the seniority and benefits that you build up. For instance, as you work more years at a company you may be awarded more weeks vacation time per year. Staying with the company can also be important in regard to stock plans and pensions.

### Coursework and Training

When you are employed, one of the great advantages, of course, is company-sponsored courses or company-paid tuition for courses that you arrange yourself. Again, this is a policy that varies greatly from company to company; it may also vary according to the educational institutions that are available.

If you take a course at a college or university, managers will probably expect you to do so on your own time. You also may have to justify how the course will help you in your current job, since your manager will probably have to authorize the tuition payment.

If this is a problem, you should be creative in drawing a connection between the course and your job. If you are working as a secretary you may say that a course on computer programming would help you to understand the problems and needs of the programmers in your department with whom you deal daily. Or you may argue that improving your general level of understanding of computers will help you in your job (if you work for a computer company).

It may also be helpful to set up an educational plan or articulate a goal that you are pursuing. A conscientious manager will be responsive to such goals. He or she may not want to lose you (if your course leads you to taking a new job), but he must recognize that if you feel stuck in your present job and want to advance, you are bound to leave anyway.

Some of the same arguments also apply to coursework that the company sponsors. This coursework may be given on company time and the cost may be charged to the manager's budget. Again, the manager should realize that almost any coursework will make you a more informed and better employee.

A clear plan and direction in your coursework will lead to a better job. The advantage of having obtained the entry-level job and being within the company is that you have a job that will sustain you, and you can take coursework to accumulate an academic background that will look well on your résumé, especially when combined with your job experience.

### Company-Run Training Programs

An entirely different type of training occurs when you enter a company-run training program. In that case you are leaving your present job and training (at company expense) for a new job. Often these programs are many months in duration; the training is usually of good quality; and you have an assured job in a new field (if you successfully complete the training). One problem is that you must give up your present job before entering the training program.

Following are examples of such programs:

- One program was a month-long course of training in computer programming. A large company ran the program several times, each time training a hundred or more relatively unskilled employees.
- A nine-month sales training program trained employees of the company who had had as few as two courses on computers.

- Technical writers for a new product line were given six months' training on the ins and outs of the computer so that they could write customer software manuals on it.

Again, the point to emphasize is that once you are in the entry-level job, the road becomes clearer and the opportunities are much greater.

# PART II

# 5

# *Which Job Is Best for You?*

This section discusses several types of high-tech jobs. It should help you to determine the jobs most interesting to you and for which you have some qualification. If you already have a good idea of what kind of job you are pursuing, you may want to move on to the next section.

This part of the book deals with the exploration stage of getting a job. You should be open to considering types of jobs that you may not have previously considered. Your goal should be broad—basically, to gain additional information about jobs that seem interesting to you. You will also want to find out whether the qualifications can be realistically obtained, based on the time, energy, and money you can invest in developing your background.

### Find a Job That Interests You

The requirements and background for each of the jobs that follow are different. You need to look at each one to see how it might fit your needs and interests. If you have a background in a particular job and like the work, you may find a high-tech job that is closely related. If so, it will require less work on your part to get the job, and you will be likely to start with a higher salary. It might even be possible to avoid the onus of starting in an entry-level job if you can convince the employer that your background is directly relevant.

If you are interested in a particular job, there are things you can do to get a better idea of what it is like. For computer programming, for instance, you can go to a bookstore and find an inexpensive paperback that will give you an idea of the work.

If you have friends in high-tech jobs, they are a valuable source of information. Do not be afraid to ask them; they are often the best source, especially in regard to working conditions and possibilities in your part of the country.

Taking a course is another way to learn more, but you should be seriously interested in the area before going that far.

Sometimes going on interviews can give you a better idea of what a job is like and what background interviewers want. Of course, you need a strong enough résumé to get interviews.

### Job Agencies and Job Counseling

If you are really at a loss as to what kind of job you want to pursue or what career direction you want to take, a job agency or job counseling agency may be good for you.

Sometimes personnel people can be helpful, even if it is immediately clear that you do not qualify for a job. Also, you may be able to extend a phone conversation with a personnel person and gain some useful information, such as the qualifications wanted and whether other jobs may be open in the future.

Job agencies can sometimes be helpful even if you do not have much in the way of credentials. I am talking about the kind of agency whose fee is paid by the employer. These agencies may not try very hard for you if you are in the entry-level category, but they may be willing to give some advice over the phone or in an interview. Agencies are conscious of their public image and know that you may remember them in several years when you have gained experience and are looking for another job. If an agency is willing to interview you, always make sure it will not charge you a fee (unless you want to work with that kind of agency).

Some job counseling firms or career programs do charge a fee. The fee can be quite substantial, although a reputable firm will provide you with substantial services in return.

However, with perseverance and work you can probably do much of this career evaluation on your own. If you are really at a loss about what to do with your job situation and want more direction, you may try several sessions with an experienced career and job counselor. Choose a counselor who focuses on career issues rather than general issues.

If you do consider job counseling, remember that there are many well-qualified counselors with master's degrees (rather than the PhD of the licensed psychologist or the MD of the psychiatrist) who can give you help at a lower cost. If possible, check out his or her reputation first. If you feel uncomfortable with someone, try another counselor. Remember that you are paying the fee, and the service and the person you are dealing with should be what you want.

Some job counselors or counseling firms offer weekend sessions for groups of people. This kind of program can be very helpful in assessing your strengths and weaknesses and interests. The cost may be a few hundred dollars. Try to check out a program by talking to people who have taken it.

You can also obtain advice from one of numerous career books available in paperback; several are listed in the reading list. One that is highly recommended to me by counselors is *What Color Is Your Parachute?* Career counselors and career programs have different approaches, and specific exercises may be different. The basic knowledge is similar, and I try to provide much of it in this book. If you buy one of these books and conscientiously work through most of the exercises, you can probably accomplish what might cost much more through a professional program.

Of course, there are other sources of help and advice. One is the job placement center of your college. The setup of these centers varies, but they usually help graduates who have been out for several years as well as recent graduates. Most have job listings, and many are willing to provide job counseling (for a few sessions) to their graduates. These counselors and offices can be valuable since they tend to know about the job situation in your area for people with your background (a degree from that specific college).

Contacts with close friends and people you know in the high-tech industry can be very valuable. Do not be afraid to take up someone's time. Most people are willing to talk and like to feel that they can help someone. Older or more experienced people may recognize in you something of themselves when they were younger and looking for their first job.

You should also realize that some of the most useful "counseling" comes from talking with your friends or your spouse in an informal way. Counselors recognize the simple value of listening to another person and providing a forum for someone to air thoughts or feelings. We all also tend to get stuck in our own viewpoints of the world and settle into certain thought patterns. Hearing a different viewpoint and the reactions of your friends to your goal or plans is helpful simply because it is a different viewpoint and may give you a new perspective or suggest other possibilities.

## Sales

Sales or marketing jobs require that you know something about what you are selling. You will have to take coursework at some point and gain the technical expertise to be knowledgeable.

On the other hand, that knowledge need not be of the deepest or most complex kind. Often the main thing to know is how to assess a customer's needs so as to recommend the right computer or right configuration of a particular computer. Many times a small variation in a computer (the amount of main memory included, for instance) can make a big difference in how useful it is to the customer. Some of the factors that must be assessed are:

- Size of the operation. This is usually related to the size of the company, although you may be dealing with a department within a large company. The overall amount of computing to be done relates directly to the size of the computer, such as how much main memory is needed, and also to the peripherals needed, such as additional memory in disk or magnetic tape form.
- Type of operation. Is the operation mostly word processing or data processing, or is there a need for both?
- Special requirements. Other issues may come up such as the familiarity of the personnel with computers, the company's compatibility with other computers, or the ability to hook into a certain type of computer network.

Coursework or training would help you to understand these matters clearly, but on-the-job experience will quickly give you an idea of how to assess a situation and what is most important for a particular customer.

Of course, one of the basic requirements for sales is a pleasant and enthusiastic personality. Many people seem to have a knack for sales and for getting along well with people. This is still important in the computer industry; although understanding of the technical side of the product is more important than in other industries, that can always be learned.

A "sales" personality is not easily learned and is more a question of what type of person you are. You probably know whether you have a friendly outgoing personality, or a more reserved one. You also need to have a positive attitude and not be easily discouraged.

There are several ways to get into a high-tech sales job:

- Having a sales background in another area. In this case you need to show a strong performance in sales and convince the interviewer that you can learn the technical knowledge needed.
- Having a college degree in marketing. You would need to take courses in computer science, but if you do, the degree would be an attractive credential.
- Getting into sales from another job within the company. You would need to show a strong interest in and aptitude for sales. You may already have gained enough technical background from your present job or from company-sponsored coursework.

For an interview for a sales job, remember to be positive and show your enthusiasm and the social side of yourself. The interviewer will be looking at how you present yourself as well as other factors. In other kinds of interviews, such as for engineering and programming, your personality and how you present yourself are much less important.

### Teaching

There are many opportunities for teaching and instructing in the high-tech field. Many companies have in-house training programs and programs to teach customers how to use the company's computers. Topics that might be covered include programming, word processing, and computer concepts.

Teaching these courses does require understanding of your topic. There is no need to understand every aspect of computer operation, but you do need to know the topic you are teaching.

Generally, there are two main ways to move into a teaching job in the high-tech industry:

- Coming from a teaching background, you need to acquire technical knowledge of an area, which may be done either by coursework or by working in the area.
- Without having a teaching background, you can work in a technical area and, after you acquire expertise, apply for a job as an instructor in that area. You then learn teaching on the job.

A big problem for former teachers is lack of a technical background. In many parts of the country, because of cutbacks in education budgets, many former teachers are looking for other jobs. This may be why teaching jobs in the high-tech industry are rarely advertised. Managers may be deluged with résumés from the many competent teachers who lack the appropriate technical background.

My advice to you if you are a teacher is to get some technical background (perhaps through night courses) and then look for a high-tech job. If you are lucky, you may be able to move directly into teaching, but do not be discouraged if you have to take a nonteaching job. There will probably be an opportunity after a year or two of experience to put your teaching background to work.

If you do not have a teaching background but are thinking of trying teaching, you will probably have trouble moving directly into a teaching job. Find a high-tech job and develop an area of expertise (such as programming or word processing). You may then be able to move into an instructing job in that area.

The requirements for a teaching job in a high-tech company are much the same as for other kinds of teaching. You must be able to lecture to a group of people without being nervous; you must have a good presentation and make the material interesting; you must evaluate students; and you must deal with administrative and logistical details.

Often in training courses in high-tech industries each student has his own terminal and operates it to do exercises during class. You must understand how the terminals work and take care of minor problems that arise with them or with a student using one incorrectly. This familiarity with terminals is not difficult to acquire.

### Personnel

Personnel is an attractive field for many people because it involves working with people. You need a general familiarity with the computer industry as well as with the kinds of jobs in it. In some situations, it may help to gain some technical background; for instance, you might have to talk intelligently with managers about what a job involves technically and what kind of background is needed to do it properly.

There is great opportunity in the industry for technical recruiters, people who bring in new employees for a company in highly competitive areas such as electrical engineering or software engineering. Recruiters seek out workers in such areas and sell them on going to work for a particular company.

The technical recruiter job requires some of the characteristics of a sales job: being affable, outgoing, personable. It may be a possibility for people who have a background in sales.

### High-Tech Salaries

Salaries vary greatly in the high-tech industry. Jobs such as computer operation, word processing, technical illustration, manufacturing work, and layout design offer salaries up to $20,000 and higher.

Other jobs such as technical writer and programming may begin around $20,000 and reach as high as $40,000 or more.

Engineering jobs pay the best. Electrical engineers may earn about $30,000 to start. Brilliant engineers or those in management may earn $60,000 or more.

| Background | High-Tech Job | Comment |
|---|---|---|
| Drafting | Drafting | Uses computer graphics video equipment |
| | Printed-circuit board layout | Uses CAD (computer-aided design) video equipment |
| Assembly, manufacturing | Manufacturing | Electronics assembly |
| TV repair, electronics | Electronics technician | Assembly line electronics work; may also work in field service |

| Machine operator | Manufacturing machine operator | Requires familiarity with terminals, since many machines are computer-driven |
|---|---|---|
| Typing | Word processing | Needs a basic word processing course |
| Machine repair | Field service | Needs some electronics background |
| Sales | Sales | Needs to be familiar with the company's product line |
| | Technical recruiter | Recruits engineers, programmers for company |
| Personnel | Personnel | Needs to be familiar with high-tech job situations |
| Art, commercial illustration | Technical illustrator | Needs general familiarity with computer equipment |
| Writing | Technical writer | Needs to be a disciplined writer and a quick learner |
| Teaching | Teaching | Needs good understanding of a technical area |

# 6

# *Computer Programming*

For many, computer programming is the golden pathway into the high-tech industry. It is something that most people can learn; it has a high and continuing demand; and it does not require a college degree.

Many kinds of programming courses are given throughout the country, at small evening technical schools as well as large universities. Companies also run in-house programming training courses or promote people from less skilled jobs.

### Hardware Versus Software

Computers can be viewed in two ways: as hardware or software. Hardware refers to the actual physical, mechanical piece of equipment that is the computer. It includes such peripheral equipment as terminals and printers, as well as a central part of the computer called a central processing unit. In very small computers (such as personal computers) the terminal and central processing unit are combined in one package, but in almost all cases the printer is a separate unit.

The hardware contains great computing power and great capacity to do calculations, word processing, or other computer functions. However, this computing power remains dormant until a person tells the computer what to do.

And here we come to the crux of the matter. You may have heard this before, and you will hear it again: The computer is stupid. It will do only what it is told to do; if the instructions given to it are wrong, the computer will give a wrong result. It does not evaluate what you tell it to do. It simply follows the instructions faithfully. If you tell it to extract the square root of 2 one thousand times, the computer will do so even if there is no point to it except to exercise its circuits.

The instructions are, of course, the software, which is the same thing as the computer programming. The instructions, or computer programs, can vary from one minute to the next. Thus, one computer (or one piece of hardware) can be made to perform many different tasks by using different programs.

The basic aspects of programming can be stated as follows:

- You must be specific and careful in writing instructions for the program. Any mistake will be faithfully followed by the computer, causing a botched result. In other words, it helps to be meticulous and a bit obsessive.
- You must be able to analyze a problem and break it down into a series of steps, which then become the instructions given to the computer. This is a kind of problem-solving and analytical ability that most people have to some degree.
- Since there cannot be any minor or typographical errors in the computer program, you must have the patience to analyze any errors (or bugs) that appear when the program is first run and to correct (or debug) the program. The complexity of many programming problems also demands

this kind of debugging to ensure that a program is working exactly as planned.

Since new programs and software can be developed to give existing computers new applications and new abilities, there would always be a demand for computer programmers even if no new computers were built.

### Programming Languages

The series of instructions that make up a program usually must be written in a specific format using a specific vocabulary. This is called a computer language, and it is much simpler than a foreign language such as French. A computer language may involve a vocabulary of one hundred words or phrases, whereas a foreign language would require the mastering of thousands of words.

You can see that a computer language is much easier to learn than a foreign language, but it can still be a challenging task and not one to be taken for granted.

Different computer languages are used for different purposes. They may look quite different. The most basic is machine-level language, which is nothing more than a series of zeros and ones (referred to as bits):

0101  0111  0100  1010  1110  0110  1100

This is the language that computers most readily understand, because it is the way the computer is designed to operate. If there is an electrical current in a circuit, it indicates a 1 bit; if there is no current, it indicates a 0 bit. The combination of many circuits and switches arranged in a complex fashion enables complicated calculations and decisions to be made based on the simple information of current or no current.

Very few programmers actually use such a basic

language to program. They use languages that look more like English or mathematical equations. These "higher-level languages" are changed into the simpler machine language level automatically by the computer.

The assembly language looks a little more like simple English:

CLA
ADD A
ADD B
STO C.

However, it still needs quite a bit of interpretation. This simple program says to clear out any previous information (CLA), add A and B together, and assign C the value of the result. In other words, it represents the simple math equation: $C = A + B$. Programmers do use this for certain types of applications, although it takes longer to write this kind of program than to use a higher-level language.

The higher-level languages are easier for the programmer to use because they come closer to what we already understand. In the FORTRAN language, one often used for scientific and mathematical applications, the above statement would be written:

$C = A + B$.

In COBOL, a language often used for business applications, it might be written:

COMPUTE $C = A + B$.

The higher-level languages can take more computer time, however, because the computer must change the language first into an assembly language and then into the machine-level language.

## Analysis and Program Writing

To write a program, then, the problem must be reduced to a series of statements that the computer can understand. This is the analytical part of programming. You might compare it to telling someone from another planet how to fix a flat tire. The alien being might be quite intelligent, but never having seen a car, or a tire, or a jack, would have no idea what to do.

You might say at first:

- Jack up the car.
- Remove the old tire.
- Put on the new tire.
- Let the car down again.

You would soon realize that this was far from adequate. You would have to make your instructions even simpler. For the first one, jack up the car, you might have to say:

- Remove the jack from the trunk.
- Put on the parking brake.
- Make sure the jack is in "down" position.
- Place the jack under the bumper.
- Jack up the car.

Even that might not be enough, and you might have to consider defining and explaining things in more detail.

But enough of our alien friend. This is similar to the process that occurs when a programmer analyzes a problem. The problem must be broken down into a series of simple instructions that the computer can understand.

Many people find this process too detailed and tedious. Any error in the instructions that does not follow the specific format of the computer language can cause the computer to reject the program. The error must be found and corrected: the program must be "debugged," and that can take longer than it took to analyze the problem and write the program.

Debugging is easier than it was in the past, and that makes the programmer's job easier. In the past you had to submit your program in a batch mode, meaning that the program was run with several other programs, one at a time in sequence. Thus you did not get your results back right away and may have had to wait a day or two.

Most modern computers run programs as you submit them. Also, error messages have been made more helpful and easier to understand, as the industry realizes the importance of the debugging process and in general seeks to make computers more accessible to the user. Some computers even examine each programming instruction as it is typed onto a terminal and tell you if there is a bug in that particular instruction.

However, it is very gratifying when a program does run correctly. You have a great feeling of accomplishment and also of power when the computer is able to digest a mass of data, manipulate it, perform calculations, and then give out the desired results rapidly.

If you are working with computer graphics, the results can be very pleasing indeed. Instead of a printout of numbers or dull accounting reports, you receive an impressive full-color display or plotted printout on a terminal. The general area of computer visuals and computer graphics is a booming one, and most newer computers have software for graphics.

### Prerequisites and Training

Many people have taken one or two courses in programming and then gone on to have a successful and well-paid career as a computer programmer. Even so, much of what I have said about entry-level jobs applies to programming as well. It can be hard to get into the field even if you have taken the appropriate coursework. I know of one situation offering entry-level training for computer programming in which there were fifty applicants for each opening. That may explain why entry-level or training positions in programming are rarely advertised and can be difficult to find.

In the past many programmers could obtain jobs without a formal degree in computer science. This will be more difficult in the future. Employers are looking for a bachelor's degree in computer science or engineering.

Any actual experience in programming or coursework can of course strengthen your résumé considerably. You will have to do some initial learning on your own. Very few companies offer entry-level jobs in programming that provide full training for complete novices.

Don't try to learn programming from a book. That would be comparable to trying to learn to drive a car or to ski by relying entirely on a book. You must have hands-on experience with a computer to understand the kinds of pitfalls and problems that can arise.

Programming languages are standardized to a greater or lesser extent; examples are BASIC, PASCAL, COBOL, FORTRAN, C, and LISP. But there are differences, often determined by which computer the language runs on. Managers often want to hire someone with experience in a specific version of a language.

Keep that in mind if you are thinking of taking a computer course. First, be sure the course offers plenty of opportunity to do actual programming. Second, be sure the computer available is a well-known brand and a recent model.

Look through the want ads in the papers and see what computers are referred to in ads for programmers. That will give you some idea of what computers are in use in your part of the country, and, for instance, whether more jobs are available in scientific or in business-oriented programming.

If you are thinking of taking a course and are wondering about the particular computer that is available, you can telephone a local sales office for the company that makes the computer. Be sure to know the exact name and model number (such as Digital Equipment Corporation VAX-11/8600). With luck, you will get someone who is willing to spend a little time telling you which of their products are most recent and most popular. At least you can determine if a particular model is of recent vintage. You will be better off doing programming on a newer-model computer.

You might also ask any friends you may have who work in high-tech jobs. They often have the best perspective on what computers are most widely used and which ones are declining or increasing in popularity.

Before undertaking to learn programming, you should study your abilities and interests carefully. Most people can learn programming, but that does not mean that most people will be good programmers or be happy in a programming career. Programming is basically an intellectual or thinking kind of work. If you are happier working with your hands, you may be happier in another kind of job. Most programming does require some mathematical ability, but in most cases a good understanding of high school algebra is all that is needed.

There are several ways to get actual experience in programming. Some of the ways are to:

· Buy a personal computer and do some programming on it, relying on the manuals that come with the computer.
· Take courses in programming at a technical school.

- Take courses in programming at an accredited college or university.
- Talk your boss into letting you do some programming on the company computer. Think up a project that will be useful to him. Volunteer to do the learning part of the project on your own time.

These approaches all have their drawbacks and benefits.

Learning on a personal or home computer allows you to learn at your own pace and to work at home. You may miss out on having someone to advise you, and employers may not be much impressed with this kind of experience.

Taking courses can look well on your résumé. You should take them at an established and well-known university if you have the opportunity and can get classes at a convenient time (such as night classes if you are working).

Try to determine if the computer is a popular model and if there are plenty of terminals so that you can do actual programming. Check to see if a terminal will be available at a time convenient for you. One large university in the Boston area has plenty of terminals, but they become flooded with night-school students at 5:00 p.m. when everybody gets off work.

Try to determine if there will be real human beings available to help you with programming problems. One-to-one help is invaluable, since some small misunderstanding on your part or misstatement on the teacher's part can cause you great difficulty in getting a program to work. Little details can cause big problems in working with computers. A small amount of advice from a teaching assistant or an experienced programmer can help a lot.

Since you must do actual programming, you will

spend a significant amount of time at the school in addition to class hours. Schedule when you will spend this time. Avoid schools that are far away, are difficult to get to, or have limited hours for terminal use.

Gaining actual on-the-job experience is one of the best ways to learn programming. Experience of this kind will enhance your résumé. It can also be difficult to get. You will have to persuade your boss to let you spend some of your time on the computer. The best way is to think up a project that can really help your boss and the company in some way. I was able to do that when I was working as a technical writer, but not for a computer company. I persuaded my manager to let me write some programs in BASIC as part of my job. The programs were not really impressive compared to what professional programmers accomplish, but they gave me valuable experience and helped me later when I applied for a job as a software technical writer.

### Kinds of Programming Jobs

There are many kinds of programming jobs and programming languages. If you are taking coursework or learning programming in another way, you should keep in mind what type of programming you are learning.

Two of the main areas are business programming, which requires a language such as COBOL, and scientific programming, which requires a language such as FORTRAN. Thus, if you decide to learn a certain language, you have in effect made a decision as to what direction your programming career might take.

Business applications require that you manipulate large files of information and be able to generate a variety of financial and business reports. COBOL is well suited for those tasks. You may spend more time updating existing data files and programs than writing new ones.

Business programming is almost universally in demand, even in small cities and towns, because more and more businesses are using small computers or buying computer services from small specialized firms. Sometimes prepackaged programs written by a manufacturer provide all that is needed, but there is still a great demand for writing programs to meet a company's specific needs.

Many people find scientific programming more interesting than business programming. FORTRAN is a language often used for scientific and engineering applications. It can handle mathematical and algebraic manipulations more easily than other languages. This type of programming job is less widely available, and you may need to be located near a major high-tech center or technical university complex.

For general-purpose programming that can also do mathematical manipulations, you can use BASIC or PASCAL. These two languages are more user-oriented and easier to learn than some of the other programming languages.

One of the newer languages that you may use is called C. In the artificial intelligence area, you may use either LISP or PROLOG. In some situations you may use a fourth-generation language that enables you to accomplish a lot with a few commands.

You can also be involved in writing programs for operating systems, the master programs that control how the computer functions. This is really a form of software engineering and requires a more sophisticated understanding of how the computer functions. Often an assembler-level language is used for the purpose. This type of language requires many more commands to accomplish the same purpose as other computer languages, so your job can be longer and more difficult. However, in a modern approach the number of commands can be cut down, for instance, by using "macro commands."

The types of background (in addition to knowing a programming language) required for these different areas vary greatly. Scientific and engineering programming may require a bachelor's degree in science or engineering. For business programming, you may need to have a degree in business or marketing and to understand finance and accounting. For more sophisticated programming, such as on operating systems, you may need a degree in computer science.

### Working Conditions and Opportunities

Programmers are often the more eccentric of technical workers. They may like odd working hours or prefer to work at home. Many companies recognize this and give programmers unusual freedom to set their own working hours. It should be emphasized, however, that the eccentric image of programmers is decreasing. So many programmers are needed that people are moving into the field from varied backgrounds.

If you are considering a programming job, find out if you will have your own terminal in your own office or whether you will have to share an office or a terminal. In most cases you should have your own terminal and be able to enter programs directly onto a computer and run them yourself.

If you have an office, it will probably have shoulder-high partitions, not floor-to-ceiling walls.

Opportunities for advancement usually arise after you have worked for a few years as a programmer. You can become a project leader or supervisor, or a systems analyst.

If you are working as a software engineer, or with advancement, your pay can move into the higher salary range. You can also earn higher pay as a contract worker, although you will not receive fringe benefits.

Of course, you can start your own consulting firm or contracting firm after several years' experience. If you can show the right expertise and ability to solve computing problems, you may be paid very well. A friend of mine started a consulting firm and soon found work with a company that distributes computers. My friend knew a particular model of computer and its software thoroughly and was soon of great help to the company; in effect, he replaced two full-time programmers. Needless to say, the company was willing to pay generously for such service.

# 7

# *Manufacturing*

Many different types of jobs are available in computer manufacturing. They do not require as extensive a technical background as other technical jobs, and some require little background at all. Many of the jobs are fairly clean and safe compared to other industrial situations, and many are not set up on a moving assembly line.

In general, these are basic industrial jobs not much different from those in other industries where machines are assembled, inspected, and tested. They involve such basic work as performing a certain operation on a printed-circuit board, inspecting boards or parts for defects, calibrating an assembled piece of equipment, or testing a piece of equipment.

### Prerequisites and Training
Many of these jobs require only a limited knowledge of electronics. For instance, as a machine operator you would need to know how to operate a specific machine and know its particular problems. This type of knowledge is usually gained on the job and for the most part cannot be learned in academic courses.

A high school education is definitely a plus, but in some cases it may not be necessary if you have good

English reading and writing skills. I have heard of government-sponsored training programs for manufacturing workers that do not require a high school education. The fact that these succeed is due to the focused nature of the training, which provides a good familiarity with electronics and is aimed at particular manufacturing jobs. These types of programs, however, are not common and are aimed at particular groups of people, usually minorities. If you have access to such a program, you should pursue it.

In most situations, however, employers are looking for a high school diploma. If you do not have one, my advice is to get one; not having it is too much of a drawback to getting a job or, if you have a job, to getting a promotion.

Specialized electronic training can be obtained in some high school vocational programs and some technical schools. Whether you really need it depends on the type of job you are aiming for or that is available in your area.

In general, you would be better off with some basic electronics training. It would be valuable to know how to operate electronic devices such as an ohmmeter or an oscilloscope. You should be able to recognize basic components such as resistors, capacitors, transistors, and chips and to know the main features of a printed-circuit board. It would also be helpful to be able to use a soldering iron and to read a simple wiring diagram.

You should also have some familiarity with use of a computer terminal. More and more machines in the manufacturing process are controlled by computers. This is often called numerical control (NC) operation. Usually there is a limited number of commands to learn, and the more complicated procedures with the terminal are handled by an experienced operator or technician. However, having some of this electronics background may land you a job that pays better and is more interesting.

### Kinds of Jobs

**Calibration** is a manufacturing job. Calibrating an electronic device after it is assembled requires using an oscilloscope or electronic meters. Each device requires individual tuning. Typically, calibration involves taking an assembled piece of equipment, putting it through certain tests, and adjusting the circuitry so that when you do the test certain values or results occur on the meters. The adjustment may be a simple operation such as turning a certain screw. You may be expected to calibrate so many devices per day or week, but a time limit cannot be set on each device because some can be calibrated quickly and others take longer.

**Quality inspection** is another major area. These jobs often require close attention to detail and a high level of conscientiousness. Such jobs are also clean and are less time-dependent since each inspection must be done carefully. They do not necessarily require an electronics background. Obviously there are certain defects or problems to look out for, but those vary from situation to situation and are learned on the job.

Quality control is very important. Computers are becoming more and more complex, and the quality and reliability are based on good engineering but also require good inspection procedures.

For instance, you might have to inspect a printed-circuit board for defects before the chips, capacitors, and other components were soldered onto it. You would have to examine small lines on the board (which are electrical circuits) for breaks or bridges between the lines, and also look for other defects. In most cases you would not do this with the unaided eye, but would use a microscope or a magnifying device. The boards would

not come to you on an assembly line but would be brought to you in batches.

**Machine operating** is another major manufacturing area. Less and less work is being done by hand and more by machines, which are often controlled by small computers. In operating these machines you would set them up to do a particular task, watch them to make sure they worked properly, and load parts or materials into them.

This work is usually clean and relatively safe. It can be quite repetitious, but despite that, you need to be alert to possible problems.

One simple machine is a mechanical scrubber, a device that cleans printed-circuit boards before they are exposed to a photographic and etching process. The machine must be set to a certain cleaning pressure (which may or may not be done by the operator); then printed-circuit boards are fed into one end and unloaded from the other end.

A more complex machine is an automatic drilling machine that drills holes in printed-cicuit boards. You would have to set up the machine very carefully and check it out to see that it was functioning correctly. Errors of a few thousandths of an inch could ruin the product. You would then load the boards onto the machine, after which the actual drilling would be controlled by a small computer.

After gaining some experience as a machine operator, you might be promoted to the position of process technician. You would then make more complex decisions and be involved in setting up the machinery.

### Working Conditions and Opportunities
As in many manufacturing situations, machine operation usually has specific performance goals to be met. However, most operations are not on an assembly line.

Operations such as inspection require close attention to detail and concern for quality.

Pay is usually on an hourly basis. It is usually good in manufacturing work but varies from company to company and from job to job. Overtime is often available.

Most operations are clean, but some are not. The latter include wet process procedures such as the developing and etching operations that occur in making printed-circuit boards. These are largely chemical processes in which baskets containing boards are moved from one tub to another. The tubs contain different chemicals, including strongly acidic and alkaline solutions that can cause burns. However, companies do make efforts to protect workers, and this is a more safety-conscious age than, say, fifty years ago. Special clothing such as plastic eye shields, rubber gloves, and rubber aprons are usually provided in the more dangerous areas. Emergency water showers and eye-rinse devices are also available.

Some operations actually require that *you* be super clean. In chip manufacturing, for instance, you would probably be required to wear special clothing and coverings over your hair and shoes and to be conscientious in clean-room procedures.

Opportunities to advance are available, including advancing to supervisor or technician. The prerequisites are usually several years' experience and showing that you are a good and intelligent worker. Coursework can be helpful, especially if you are interested in a supervisor's position. Some companies have in-house training programs for potential supervisors, and many colleges offer managerial courses.

You can also advance from the manufacturing area into office work, although that is harder to do. Some companies offer training programs. Another route is obtaining a college degree through night school. That can take time and be quite a strain, since you would be working full time during the day, but if you can stay with a degree program and complete it, it can be well worth while.

# 8

# Drafting and Layout Design

The old image of the draftsman hunched over a drawing board with pen and ink is giving way to that of the draftsman sitting in front of a sophisticated computer system using a keyboard, a light pen, and a drawing tablet. Modern computer graphics systems and computer-aided design (CAD) systems are becoming more widely used. They are used by engineers, and they are also used by drafters to make drawings of machine parts. Even fashion designers are using CAD to assist in designing fabrics.

The systems are becoming much more sophisticated and now are available in color. They may show several layers of a drawing overlaid on each other. The most advanced systems produce impressive and sometimes very realistic pictures. Some of the most sophisticated results have gone into computer animation such as that seen in movies like "Tron" and "Beauty and the Beast." Directors continue to turn to computers to assist with animation and special effects.

Much of this advance is the result of the growing power available in small computers. At first, CAD software for mechanical or electrical design projects was intended to run on mainframe computers. Next, mid-range workstations were developed for graphics software.

Now, with the development of UNIX—a multitasking operating system—CAD software can be run on PCs such as the Apple Computer Inc. Macintosh.

### Layout Design

A job that is new with the development of computers and advanced electronics is that of electronic parts layout. The parts are printed-circuit boards or integrated circuit chips. The job title is layout designer.

A design is created by an engineer, for example, for a printed-circuit board. The design specifies certain components that will go onto the board and the connections that must be made between those components. However, the engineer does not specify the exact routes that the connections are to take on the board. A printed-circuit board layout designer does that, usually relying on a CAD system.

The process can be quite complex. Just as more and more circuitry is being crammed into each chip, printed-circuit boards have become more complex as more and more chips and other components are put onto each board. So many connections are required between the components that routing the connections becomes a problem because they get in each other's way. The placement of components and the routing of connections becomes a puzzle to be solved, and you need analytical ability to lay out the board. It also helps to have a good spatial sense and good visualization.

Fortunately computer programs have been developed that help in the routing problem. A program can be run and the routing results be displayed on the designer's video screen. The designer is still needed, for rarely does the program do a perfect job. The designer still must route those connections that the routing program could not route and check the program's results, since the program may create routes that seem illogical to the human eye, and the designer can easily improve on them.

The chip layout designers perform a similar task, except that they are trying to fit into a chip a certain pattern of circuitry designed by an engineer. There are no components to position in the chip, but there may be routing problems between different areas of the chip.

A good designer, working with good CAD equipment, can fit more circuitry into a given amount of space. What this ultimately means is that the computer can be smaller and more powerful. That is the trend in the computer industry—to be smaller, more powerful, and cheaper. The layout designer thus works (along with engineers) at the leading edge of the advancement of computers.

For these reasons, this can be an exciting area in which to work. Also, working with graphics is fascinating. Much computing work involves manipulating numbers or, as in word processing, text. CAD systems and computer drafting systems give you a visual result that is less abstract and more interesting.

### Prerequisites and Training

A drafting background is a basic requirement in these areas. That background can be obtained in a technical school. An architectural background or architectural, drafting, or drawing courses in college can also be useful. The general requirement is that you be able to think visually and be comfortable working in a visual mode.

A knowledge of electronics can also be helpful, but extensive knowledge is not necessary. In drafting work such as drawings of parts, computer layouts, or building layouts, many of the requirements are the same as for other kinds of drafting.

The difference is that you would be using a computer graphics system. A job interviewer will look for familiarity with such a system and sometimes for familiarity with a specific type of system. Other kinds of background in electronics or programming may be helpful. Some of the same kind of thinking involved in programming is used when operating a computer graphics system. In a sense, you are programming the graphics system to produce a specific drawing.

Unfortunately, not many schools offer courses in using computer graphics systems, and you may have to rely on a company providing training on a specific system. Of course, if you can find such a course, pursue it. If you do become familiar with one type of computer graphics or drawing system, it will help you in learning to use other computer drawing systems.

For circuitry layout, many of the requirements are much the same as for drafting and computer graphics. You do not need much background in electronics. That is because any important electronic decisions about the layout will be referred to the engineer who designed the circuits. Also, basic requirements about the circuits are specified beforehand, including such things as spacing between connections and critical placement of certain components of the layout.

There are many types of layout systems. Some large companies devise their own systems for internal use, whereas others make systems and market them to other companies. Job interviewers sometimes look for specific experience on the system that their company uses. If you have a choice you may want to learn to use one of the commercially available layout systems (rather than one that is used at only one company); that would make it easier for you to change jobs at some point in the future.

Few schools offer courses on circuitry layout, so the best background is in drafting, some electronics, and some programming. A college degree is not necessary.

**67**

Circuitry does require analytical and puzzle-solving ability, though it is difficult to cite a specific background for those skills. You also need to be meticulous about details: A small error can ruin the whole board.

### Kinds of Jobs

Drafting jobs in high-tech are quite similar to other kinds of drafting jobs except that the drawings are of computer parts and printed-circuit boards.

Circuitry layout can involve printed-circuit board layout, which requires assigning positions on the board for the components (such as chips) and then indicating the routes for the connections between them. The components have small leads that fit into holes drilled into the boards. The connections are actually made from hole to hole and are represented on the finished board by a thin layer of copper that has been etched to the specified pattern of connections.

You thus become concerned with such parameters as components, hole sizes, connection patterns, and connection widths, all of which must be specified accurately. Also specific requirements, such as a minimum connector width, must be followed for these parameters.

Circuitry layout is also done at the chip level. This is in many ways similar for the designer, although the result is much more microscopic. There are no discrete components, but instead certain discrete patterns of circuitry. These patterns are called *gates*, which can be thought of as an elaborate electronic switch with connections between separate gates or groups of gates. Basic criteria exist for the width of connections and for the placement of certain kinds of gates.

Other kinds of layout involve other kinds of computer parts. These include *backplanes*, which are large printed-circuit boards into which smaller boards are plugged, and various newly developed chips that have different layout requirements from the older kinds of chips.

In general, you would work at the chip level or the board level. After gaining experience in one area, it may be difficult to switch to another kind of layout later in your career. It may be argued that chip-level layout is more secure in the long run, since printed-circuit boards may become less important as most of the important circuitry is placed in chips. In some areas, such as printed-circuit board power supplies, there will probably always be boards, because the high electrical power requirements would burn out the small circuits of chips. In my own view, printed-circuit boards will remain important because only in the smallest computers will the whole computer be placed on the chip. Larger computers will still need the boards to tie together the chips that make up the computer.

### Working Conditions and Opportunities

Since you would be working at a CAD station, circuitry layout is clean office work. Using a CAD station avoids the need for a multiplicity of inks, pens, and drawing or lettering templates. Some people may actually miss these and wish to return to being able to draw things directly, by hand. The great advantage of a CAD system or computer drawing station is that corrections or changes in the drawing or design can be made easily and without erasing or whiting out parts of a drawing.

An important concern is the quality and speed of the work station. Some CAD or drawing systems are connected to larger computers along with many other stations. If all the stations are in use, you may not be able to change the drawing very quickly. When you give the computer a command, it may be minutes before the system responds with the appropriate change in the drawing. You may be better off if you work at one station supported by one small computer.

While on one hand more power is being packed into computers, which should decrease the response time, on the other hand more is being demanded of the systems. The trend is toward color systems, which place greater demands on the computer.

If you are being seriously considered for a drafting or layout job, you should be able to ask for a demonstration of the equipment you would use. In general, a color system indicates a more modern system. Observe the response time on the system, whether it is a matter of ten seconds or so or more like a minute or minutes. Certain operations do take a long time (in minutes), such as redrawing a complicated layout in color on the display screen. The thing to notice is whether the system is continuously working on the drawing—adding parts and outlines—or whether it hesitates or stops periodically. Hesitation may indicate an overloaded system, which would be frustrating to work with.

More modern systems have freestanding workstations that occasionally rely on a separate minicomputer or mainframe for special computing problems. The workstation can be freestanding because it contains more computing power. A remote minicomputer may be overloaded at times and slower to respond than a free-standing station.

Typically, systems have different ways of entering messages and commands to the computer. One is an ordinary computer keyboard onto which certain commands are typed directly. A second keyboard contains numbered keys (for instance, from 1 to 20), each key representing a specific command. A third device is a flat tablet that can be written on with a special pen. Commands can be entered this way, and in some cases you can draw directly with the pen on the tablet while what you draw shows up on the screen. A fourth device is a light pen that allows you to point to a part of the design on the video display to indicate some action to be done to that part (such as coloring it a certain color).

Still another entry device is the *mouse*. This is a small round or square wheeled device with buttons. By moving the mouse and pressing the buttons you can select different parts of the screen or different functions.

In general, you should look at a system to see if it seems comfortable to use, if the display screens and keyboards are conveniently positioned, if you will have a comfortable chair, and if the room is darkened. This is important since you will spend much of your time sitting in front of the system.

Opportunities exist for advancement into supervisory positions, as well as into senior layout or training.

Very good wages can be made if you work on a contract basis, but you would need to be familiar with several widely used systems so as to be able to work at different companies. Wages are on an hourly basis, which may actually be to your benefit, since rush orders may periodically come through that demand extra hours of work.

# 9

# *Field Service*

Field service is repairing computers and related equipment. It requires a basic aptitude for working with and repairing machines, some understanding of electronics, and a willingness to travel to the sites where the computers are installed.

However, it does not require a high level of electronic skill, and it can be an area of great opportunity for advancement. Some surveys list field service as one of the fastest growing in new job openings and opportunities, which can make it easier to enter the area.

There is a fundamental reason for this high level of opportunity: As more and more computers are made, more and more service personnel are needed to repair them. If a company sells 2,000 computers a month, it needs a manufacturing plant that can make that many computers. The company will have to sell more computers (for instance, 2,500 a month) in order to enlarge its plant and hire more manufacturing workers. However, even if the company does not increase its production but continues to sell 2,000 a month, it will still have to hire more service personnel because each month 2,000 more computers will be installed that will need servicing. The assumption here is that many of these computers will be in new installations.

Even though computers are fairly reliable, they do need repair just like any other mechanical or electrical device. The repair work is not as difficult as you might think. Often all that is needed is replacement of a printed-circuit board. That is easy because most boards can be plugged into a socket in about two seconds.

Tracking down which circuit board is causing the problem is more difficult, but again not as difficult as it may sound. Often the approach is by formula: If such and such a problem is occurring, then probably a certain printed-circuit board is at fault. The repair strategy is to replace that board and see if the problem disappears. Service people sometimes carry around briefcases full of printed-circuit boards just for that purpose.

Field service work can be quite satisfying. When a computer system breaks down, the work of an office can come to a standstill. You may be regarded as a kind of savior who can bring everything back from stagnation to effective operation. You can have a great feeling of power and accomplishment in bringing a system back to life.

It can also be frustrating, however, when you confront a difficult problem or the people who called you in are impatient.

### Prerequisites and Training

Interviewers often look for a background in computer science and electronics such as a degree in the field or a course of study at a technical school. It may be possible to enter the field from some other background, such as experience with computers from some other kind of job. You may have to convince the interviewer that you have an aptitude for the work, but you probably will need some formal coursework or a training program.

My own opinion is that if you have a background in

repairing mechanical or electromechanical devices and have an aptitude for it, you would do well in this field. Many people have had some informal experience such as making minor repairs on their car. You need manual dexterity and aptitude for working on machines. Though many repairs require the simple printed-circuit board replacement described above, at times you will be required to do more mechanical repairs, such as replacing a disk drive or various fans and blowers. This is the traditional nuts-and-bolts kind of repair work, and usually the source of the problem is easier to locate —it is obvious, for instance, if a blower is not working.

You need to know something about operating a computer, such as entering commands or running software programs. This is necessary so that you can see for yourself what the computer is doing wrong. In some ways, however, it is secondary to the basic electronics knowledge, since you need to know only a limited number of commands and operations. You also need to know the specific model of computer you will be working on; you would be trained for this by the company, or at least manuals would be available.

Repairs may also involve use of a soldering iron. Soldering is usually avoided in favor of replacing the whole printed-circuit board and returning it to a service center to be repaired. But there may be times when you need to use a soldering iron. That skill can be acquired through courses at a technical school or by buying a kit for an electronics device and practicing soldering as you build the device.

### Kinds of Jobs
Field service jobs can involve working on various kinds of equipment, including the central part of the computer, peripherals such as disk drives, and specialized computers such as word processors.

Peripherals tend to be more electromechanical in nature. If you like the nuts-and-bolts side of the work, you may be more interested in this kind of work. Just as printed-circuit boards are replaced as units, on the more mechanical side of computers parts such as disk drives may be replaced as a unit. This modular approach to repair is being more widely used. The replacement can be accomplished quickly and the whole unit returned to a repair site for testing and repairing. The unit may then be used as a replacement part after it has been tested.

The central processing unit of the computer is the heavy-duty guts of the beast and requires more of a basic understanding of how the computer functions. This repair work is more likely to be of the printed-circuit board replacement type, and you need to know something about operating the computer. By giving it various commands or by running diagnostic computer programs, you track down the exact nature and probable cause of the problem.

Installation of computers is usually handled by field service personnel. This can be a desirable type of work, since it involves staying in one place for at least a few days, rather than traveling to several sites on the same day. Periodically computers must be upgraded to a new version of the same computer. This can be similar to installation work in that you may stay in one location for a period of time. However, there can be more pressure to finish the job, since the customer may need the computer to conduct his business and can afford to have it down for only a limited time.

Another area of field service work is closely related to manufacturing: printed-circuit board repair, or rework. The printed-circuit board that is taken out of a computer must be tested, repaired if possible, and returned to the field as a replacement part. This work involves using testing equipment and identifying faulty components or broken or shorted connections on the board. Repairs can involve work with a fine-tipped soldering iron under a microscope, or drilling holes in the board with a fine-tipped milling machine. The wages in this area are not as high as for actual field service work. Although hand-eye coordination is necessary, you do not need much technical knowledge of computers to do the job.

### Working Conditions and Opportunities

In general, working conditions are fairly clean and pleasant, since most computers are located in office-type environments. In some cases you may have to work with a computer in a factory or other potentially dirty environment.

The work itself is clean; there are none of the messy oils and lubricants involved in heavy-duty machinery or automobile repair work. You may have to use a soldering iron, but that is not dangerous or messy if you are careful.

Wages are usually on an hourly basis. That can be to your benefit, especially for overtime work. In such cases you may be paid at an overtime rate for traveling to a site or may receive payment for a minimum number of hours (such as three hours) even if you work for a shorter period of time on the repair call.

Service personnel are often graded into junior and senior personnel based on experience and ability. The senior personnel are often called upon to deal with the difficult problems.

The opportunities are definitely there. Service departments must expand to handle the large number of computers that are continually being installed. New supervisors must be found to staff these departments, so there is usually good upward mobility.

This is one area where you may have a good opportunity to establish your own business if that is your desire. You would need to work for a company for several years to gain an expert knowledge of repairing a line of computers. Once you have that knowledge, however, you should be able to run your own business repairing those particular computers.

# 10

# Computer Operating

The computer operator keeps the computer running so that the people who use terminals or do programming have it constantly available. The job includes such chores as changing tapes or changing disks, calling repair people when they are needed, taking printouts off of the printer, submitting large batch jobs to be run, and handling various administrative tasks.

The computer operator usually works with larger systems such as mainframes or the larger minicomputer systems. He or she works in the actual room with the computer and communicates with it over a terminal. Usually the actual users submit jobs to be run from their terminals; for very large jobs they give the operator stacks of punch cards (obsolete in most places) or tapes that contain the computer program and/or the data file that the program will use. Often the computer automatically schedules when jobs are to be run, but for special or large batch jobs the operator may be involved.

The job can be interesting because you are working with the actual computer hardware. The equipment is expensive, and you have to be responsible in your job. You can gain a feeling of accomplishment from being able to work with and control this large and powerful piece of machinery.

On the other hand, the work itself can at times be dull, involving such mundane tasks as putting magnetic tapes onto tape drives or taking printouts off of the line printer. The operator does not need to know computer programming and does not necessarily need to know most details of how the computer works. The job can be lonely, since you work only with yourself and the machine, and sometimes it can involve night work.

This job, however, can get you started in the field and obviously can show future employers for other jobs that you feel comfortable with and can work with computers. It is also possible to move into a job as a programmer (though that is not guaranteed).

### Prerequisites and Training

You do not need a college degree, though some schooling or familiarity with computers would help. Technical school courses give the right level of background.

Paradoxically you need not know a great deal about computers to be able to handle this type of operation. Programmers or the users will do their own programming. You need not know how to fix the computer, since you would be expected to call field service for any problems that arose (unless they were very simple). You need not know details about arithmetic logic units, or cache memories, or bus systems.

For the most part, what you do need to know is a basic set of operating commands for communicating with the system. The commands are limited and are not difficult to learn. When a large batch job is running, you will spend many hours without engaging in any communication with the computer.

The advantage is that many administrative and logistical chores are handled automatically by the computer's operating system, thus relieving you of much detailed and tedious work. The operating system automatically registers when people enter or leave the system (log in or log out), check whether they are authorized to use the system, bill their accounts, and retrieve files of data from on-line memory banks such as disks.

You need to learn some basic manual operations such as putting on new tapes or putting new disks in disk drives. These operations are not difficult, but you need either specialized coursework or training by your employer.

You need to be conscientious: The job is a responsible one. You need to be alert and responsive when action is needed from you, yet you may go through long periods when nothing much is required.

If you are interested in being an operator and are looking into technical school courses, there are some things to watch out for. Be sure that you will get hands-on experience changing tapes and disks and entering commands into the computer. Classroom work is helpful, but for this job hands-on experience is better.

You also need to know what kind of computer you will be using. Interviewers look for experience on the model of computer that they have in their company. Of course, this specific kind of experience cannot always be found, and the company must settle for less.

Nevertheless it is to your advantage to consider what kind of computer a school has available. If you find yourself trained on an obsolete or unpopular model, even if the schooling was good, you would have trouble finding a job. One way to judge is to look for big-name computers such as IBM, Digital Equipment Corporation, Control Data Corporation, Sperry Univac, or Burroughs. Another is to look through the want ads in your local newspaper. Employers usually list the model of computer for which they want an operator. By checking the ads for several weeks you can get an idea of what types of computers are being used in your area. Then you can look for a technical school with one of those computers.

## Kinds of Jobs

Computer operators work in several kinds of situations. Some work with a large computer running batch jobs. Such batch jobs involve manipulation of large amounts of data such as may be involved in mail-order business operations, Census Bureau calculations, and computer modeling applications (such as global weather models). You need not know much about the particular program except that it uses up large amounts of computer power and requires relatively little intervention by the operator. In this kind of job you may end up with plenty of free time during which you are simply watching the computer run.

In this type of job you are relatively undisturbed and may not have much to do. That could be an advantage if you have something to do with your time, such as studying for courses. However, the job can be lonely. Whether you like it depends on whether you prefer working alone or with a group of people.

Other time-sharing operations involve running a computer that has many users, who may be in the same building or may be connected by remote hookups over phone lines. This is likely to be a busier job. You would tear off printouts for users and place them in cubbyholes to be picked up. You would load tapes onto tape drives promptly after receiving a request, since a user would be waiting and doing nothing until you provided the service. Users may also have special requests, such as loading a printer with a special kind of paper.

An advantage to this job is that you deal more with people, talking on occasion with users about special jobs or problems. Some people are glad to have this kind of interaction, but if there are a lot of problems the interaction can be a nuisance. Many people prefer to have things to do in their job. If you have been in a busy job, it may sound nice to be able to sit back and simply watch a computer operate, but after a while it can become terribly boring.

A newer type of job is that of personal computer or workstation administrator. In this case you support a department of people working on personal computers or workstations. You may deal with problems of the computer itself or how it connects to a larger computer or network. The job title may be personal computer technician, network administrator, or system administrator. In this situation you deal with many people and a large number of small problems.

### Working Conditions and Opportunities

For the most part you work alone or with one or two other people in a room with a computer. You may have occasional interaction with users.

The job is clean; all you handle are paper, disks, and magnetic tapes. Most computers require air conditioning to maintain a certain temperature. In hot climates this assures you of a comfortable working situation.

A disadvantage is that many computers are operated twenty-four hours a day. That means that operators must work night shifts, either the second shift (4:00 p.m. to midnight) or the third shift (midnight to 8:00 a.m.). There may be extra pay for working the night shift, and with experience you can usually move into a day shift. If you are trying to get started in the field in an entry-level job, you will probably have to start at night.

If you work a night shift, be sure that you have quiet during the day for sleeping. Try to avoid alternate shifts such as working a night shift for a month and a day shift the next month. Your body takes a week or more to adjust to each shift, and during the adjustment (which is similar to jet lag) you will feel off base and not up to par.

Wages are on an hourly basis. There often is opportunity for overtime, since the night shifts are difficult to fill with experienced people.

There may be opportunity to advance to a supervisory position. Sometimes operators have also been able to move into computer programming positions. Companies sometimes prefer to promote an in-house person to an entry-level programming position because that person is a known quantity and because ads for entry-level programming positions often draw overwhelming numbers of replies.

# 11

## Word Processing

Word processing can be a great deal of fun for those
who have struggled with typewriters and with various
schemes of correcting errors or making changes in
documents. Most of us who have done writing or typing
have also been confronted with the chore of retyping a
page of text even though a change may affect only a
sentence or two or a paragraph.

With word processing you can avoid most of the
problems with changes. Alterations can be made in the
original text on a video screen and a fresh copy be
printed out.

Tasks like moving paragraphs from one part of a page
or document to another are fairly easy. Standard paragraphs can be printed out from the word processor's
memory, avoiding a lot of repetitious typing. Tabs and
page margins can also be changed easily, so that the
format of a document can be changed without retyping.

For instance, to move a paragraph you would take
several simple steps. You move the cursor (or blinking
light on the screen) to the beginning of a paragraph.
You then push a key to indicate that this is the beginning of the paragraph that you want. You move the
cursor to the end of the paragraph and then press a
key that indicates you want to cut out that paragraph.

The paragraph disappears (and is stored in the word processor's memory); you can then move the cursor to the location where you want it to reappear. By pressing a specific key the paragraph will reappear.

That may sound complicated, but after you have done it a few times it becomes automatic. Other functions on word processors follow a similar process. You move the cursor around and then press special keys for certain functions. Most word processing manufacturers have easy-to-use manuals and handbooks that are a great help when you are learning a particular system.

Of course, word processing does not avoid all problems. You must still perform the basic chore of typing a document when it is first entered into the word processor. In this case your job is much like that of a typist, with the word processor serving as an elaborate typewriter. It is, however, much easier to correct typographical errors.

The word processing work can be tedious, whether it is typing or editing. It can be much easier if your day is broken up with other tasks—if, for instance, your job is basically secretarial or administrative and you use word processing at times for letters, memos, and reports.

There is, however, a great need for experienced word processing personnel, and I emphasize "experienced." More and more secretaries and typists are using word processors and have some understanding of these machines. It is not very hard to get a basic grasp of how to use them, especially if you have the basic skill of typing. But to master a word processor and understand all its capabilities requires training and experience. If you reach that level, you will find that people beat a path to your desk asking for advice, and your skill will be greatly valued.

### Prerequisites and Training

This job requires English language skills; you would be expected to have basic knowledge of spelling and grammar.

It is important to remember that a word processor is a small computer. It can be confusing to people who have had no experience with terminals or computers. Still, what you need to know to operate a word processor is not complex.

Like a computer, the word processor does what you tell it to do, and you must follow the rules for telling it what to do. If you press the wrong key or give it the wrong sequence of instructions, the word processor will faithfully follow the wrong commands. For instance, if you accidentally press the wrong key, you may capitalize every letter on a page of text.

A second problem area for the beginner can be in using files. A file is merely a single document. A letter, a memo, and a report would each be in a separate file. Files must be given names, called up when you use them, and a key pressed or command given to store the file in memory. You may want to have a copy of a file to make sure it does not get lost. If you are working on a file (for instance, typing in new text) you may have to perform an update periodically (if the word processor does not do it automatically); this is necessary to make sure that what you have typed is not lost if the system "crashes."

File manipulation can be difficult for the newcomer, but it is basic to word processing and to much computer programming. Courses in word processing at a technical school or secretarial school will be most helpful. If you take such a course, it is important to get hands-on experience on a word processing terminal, for the operation cannot be learned only from a book. As you gain experience your fingers learn "where to go" on the keyboard, and the procedures for certain functions become automatic.

Most word processing systems have similar functions and can do similar things, but the keys may be arranged differently on various companies' keyboards. You will need to study the manual if you move from one company's system to another, but your basic familiarity with a word processing system will still be a great benefit

Interviewers may want word processing applicants to have skills on their particular system. As I have suggested for other jobs, you may be able to get a line on what systems or software packages are most used by checking the want ads to see which are mentioned. Word processing systems such as Xywrite and Word Perfect have become quite popular, but almost every computer company has a word processing system of one kind or another. Word processing is a fairly transferable skill. If interviewing for a job involving a machine with which you are not familiar, you may want to emphasize your word processing skills, not your skill on a particular word processor.

If you do take a course, you should get the training on a word processing system; that is, one designed specifically for word processing.

Many computers have word processing capabilities or word processing software packages. These are computer programs written for an ordinary computer that enable it to handle text more easily and do word processing functions. They are more difficult to use (though the recent ones are becoming easier) because most computers are designed to manipulate numbers and data; word processors are specifically designed to manipulate text.

You should avoid taking your initial training on such a software package, because it is much more specialized and the expertise you gain is less transferable.

The important distinction is that word processor hardware and software are both designed to handle word processing. The hardware is the machine itself, and the software is the programming that runs the machine. Computer hardware is of a more general-purpose design, and the computer is described as a computer: minicomputer, office computer, or small-business computer. The software may be designed for word processing and may be called a word processing, text-handling, or text-editing software package.

Computers today can run a variety of software packages, including word processing programs. They also run traditional business software, including accounting programs like Lotus 1-2-3. The computer industry is moving away from single-function machines and is focusing on developing multipurpose business workstations.

### Kinds of Jobs

Generally, there are two kinds of job situations for word processing people. First is the full-time word processing position, in which you would spend almost all your time typing in material (input) or editing material that is already stored on a disk.

The second situation is that of doing word processing intermixed with other chores. In most cases, this is a secretarial position that demands word processing as part of the job. This type of job can be more interesting because of the variety. Full-time word processing can become tedious; it is little different from being a full-time typist.

The full-time word processing position may pay a better salary than ordinary secretarial work (though not more than specialized secretarial work). That is because it is still a new skill and in high demand.

Whether or not you have worked before as a word processing person, you may want to inquire about the workload, whether other duties are involved (even if the job is described as a word processing job), and whether you will have opportunity for frequent breaks. Most people find it very difficult to work every minute of the day in front of a video screen. It puts a strain on the eyes and can cause headaches. One solution is to take a break, walk around, and look at distant objects (in order to focus your eyes at a distance). The video screen can be adjusted to various brightness levels, and sometimes you have the choice of white lettering on a black background or black lettering on a white background. Some companies produce screens with a green or orange text that is supposed to be more restful for the eyes.

Even if you do not have scheduled breaks, in many companies you can take five or ten minutes here or there during the day.

You may also want to ask whether the job involves mostly typing or mostly editing. It is the lengthy inputting of documents that is most tiresome. A mixture of inputting and editing is easier to deal with.

In some situations you may be expected to know grammar and spelling. If you are working for technical writers, that may not be required, but if you work for an engineer or manager he or she may rely on you to check grammar and spelling. Some word processors have automatic spelling programs that inform you if a word is misspelled.

### Working Conditions and Opportunities
This is office-type work closely related to secretarial or typing jobs. You may not have your own desk, but you should have your own word processing station. If you

are expected to do a lot of word processing and are asked to share a station, you have a problematic situation. Word processing work requires a terminal, just as typing requires a typewriter.

You may want to find out for whom you will be working, especially whether it is one person or several people. With several people you may have conflicting demands and priorities. This can be avoided if there is a clear priority system, or if someone sets the priorities for jobs as they come in.

If you are talented and know your word processor backward and forward, you will be in demand. You may be able to advance to the position of head word processing person or supervisor. These jobs make more demands, such as setting priorities for jobs or dealing with complicated formats such as tables.

Wages are on an hourly basis and generally are higher than for general secretarial work. Pay is increased if you advance to a head or senior word processing position.

# *12*

# *Technical Illustrating*

The job of technical illustrator appeals to many artistic people because they can use their talent as part of their job. This is a great advantage for talented artists who cannot support themselves doing "pure" art.

The job does not require great technical understanding of what you are drawing. In fact, you may not be expected to have any technical knowledge. Usually someone else, an engineer or a technical writer for instance, directs you as to what drawings need to be done and provides rough sketches, photos, or an opportunity to examine the object itself.

The rapid development of graphics workstations means that you may be making line drawings on a workstation. This approach may be different because you have to combine lines, arcs, and boxes to construct complex drawings. This can be learned with practice. It is not simply a mechanical process because it still requires the artist's ability and "eye" to produce a good finished drawing.

Depending on your background, you may need to make an adjustment to the corporate environment. Usually you are asked to make drawings in a certain style, and you may not have much choice about such basic factors as angle of view. You also may be asked to make minor changes in a drawing before it is considered final.

Generally, detail is very important; often drawings are relied on to convey technical information. If the drawing is wrong, then the book, manual, or brochure is wrong too.

### Prerequisites and Training

The basic prerequisite is ability to draw. This seems to be an ability you have or do not have, not one that you learn easily or with great difficulty. You do not have to be a great artist, but you must have some basic sense of sketching and drawing.

You can get a job as a technical illustrator mainly on your ability. Courses in technical illustration or commercial art or a degree in art can help, but the main credential is being able to do technical drawings and having examples of your work to show an interviewer. You do not need an elaborate portfolio, and it may not help to show nontechnical art, but two or three good technical drawings will establish your ability in the interviewer's mind.

It can be difficult to get an entry-level position, because interviewers as always look for experience. It is sometimes easier to begin with free-lance or temporary jobs, which do give you some experience and also some drawing examples.

### Kinds of Jobs

There are many kinds of illustrating: for technical manuals, for sales brochures, for popular technical magazines, for technical books, and so on.

In some situations you would have to be very much aware of deadlines, as for a technical magazine that must meet publication dates.

You may want to find out whether you will be working with a group of illustrators or be the one illustrator for a department. If you are the one illustrator, you may be caught between conflicting demands from various members of the department or may be overloaded with work. Working in a group of illustrators (such as a centralized department that does illustrations for a whole company) can spread out the workload and can provide you with the support of coworkers with a similar viewpoint and (presumably) an interest in art that extends beyond the workplace.

Much technical illustration is based on line drawings. You may be able to work in color and with more interesting and challenging illustrations if you work for a magazine. However, you would have to work where the magazine's offices were located, probably a major publishing center such as New York City or one of the major high-tech areas such as eastern Massachusetts or central California.

You may be asked to work on a graphics workstation to make drawings. These systems are usually used for drafting or design purposes, though it seems clear that they can be used for any kind of illustrating or drawing. The most up-to-date computer graphics systems can do very well as a drawing tool.

If you are expected to work on such a system, you should look at it carefully; some systems work, for instance, by assembling geometric forms such as lines, circles, and rectangles into the main outline of a drawing and then changing those lines or curves appropriately. Such a system can require quite an adjustment if you are used to drawing directly onto paper.

Some systems allow you to draw directly onto a drawing tablet, with the lines you make on the tablet appearing on the screen, but others do not give you that capability even if they have a tablet. The tablet is sometimes used to enter commands into the system, for instance by pointing a pen at a particular labeled square on the tablet.

If you are interviewing for a job that involves a graphics system, you should ask for a demonstration of the system. That may give you an idea of whether you would feel comfortable with it and whether it seems suited for illustrations.

Complex drawings can be done on such systems, but it takes a long time to add fine detail and careful work to make sure the drawing comes out right.

### Working Conditions and Opportunities

This is usually an office job paid on an hourly basis. You should have a drafting or drawing table to work on and ready access to artist's and drafting tools and supplies.

You may often work with engineers or technical writers who have requested a certain drawing. Usually you are given some kind of basic information, such as a rough sketch or photograph. These people may or may not be easy to deal with; some people, for instance, are not quite sure what they want and may request changes in the drawing after the first version is finished. Change can also be necessary if the design and appearance of a product have changed. You need to be diplomatic with people and be able to work with them.

# 13

## Technical Writing

I have enjoyed working as a technical writer and take pride in the skills I have. It can be difficult to make technical material easily understandable, but the result is gratifying when the finished product reads well and people compliment you on it. You try to be sure the information is well communicated and is friendly to the reader.

That is very important, for how well or poorly a manual is written can affect whether a product is used well. A poorly written manual can make it difficult to use the product and can prevent it from being used in the most efficient way.

Many well-educated technical personnel do not know how to write well. Your job may involve starting with written information that is adequate in the sense that it covers the necessary technical content, but does not communicate well. Your task becomes in effect to translate the material into more readable English and to avoid excessive use of "computerese" and jargon.

The people you work with are crucial to the difficulty of a project. Most technical personnel who review or provide information for documents are reasonably conscientious and helpful. Sometimes they are busy with other responsibilities (or regard documentation as a secondary responsibility); and there are always a few who are not helpful or are very slow in responding.

A basic skill is being able to get along with and work closely with people. You usually bear the burden of establishing and maintaining the working relationships. In most cases that is not so difficult a task, but sometimes it can turn into a headache.

You may start a document by receiving an assignment from an editor or supervisor. Usually you have a rough draft or notes prepared by a technical person, or the name of a person to contact to obtain information. The format of the document is usually already established because the writing department has worked on that class of documents before (such as software user guides). Each document can have its own particular variation or adjustment to the format.

You then collect the appropriate information and work up a rough draft. The draft goes through several reviews, by an editor in the writing department and by technical personnel. It goes through changes based on feedback from the reviewers until eventually it is finalized, approved, and sent on to typesetting and publishing.

The process can be lengthy, and there may be periods of no activity at all on the document. You may be working on several documents, each in a different stage of production. You need to carry on several projects at the same time and not let any one fall by the wayside.

Technical writing appeals to many people who have writing skills but have not done this type of work. Like illustrating, it can appeal to those with an artistic or a literary bent who would like to find a way to earn a living while still using their creative skills. It also involves an adjustment to the corporate environment; it requires discipline, meeting deadlines, and writing for other people (rather than for yourself).

### Prerequisites and Training

The credentials for a technical writing job are not standardized, though I can describe the general background. You need basic English language skills such as spelling and grammar. You also need to write clearly, without unnecessary flourishes or wordiness. Your style should be readable and flow well; however, you need not be a brilliant writer. What is needed is basic journeyman's ability.

Writing groups have various structures for editorial review, but in most cases an editor reviews your work to catch minor lapses in grammar, spelling, or wording.

You need to be able to work in a corporate environment and to work well with other people. You need not be an extrovert or have a salesman's personality, but you do need some social and interpersonal skills.

You need to be able to handle and understand technical material, but in most cases you need not be an expert or have a computer science degree.

In some cases you need a strong background in computer science; for instance, if you have to write a maintenance manual for the central processing unit of a computer, you need to understand the logic schematics for the computer.

But for many other situations you learn what you need to know on the job—by asking questions, by reading documents, or by taking company-sponsored courses. Many companies give month-long training programs so that you will understand a certain product thoroughly.

In general, a bachelor's degree seems to be a requirement for technical writing, but it can be in English or a nontechnical area.

Specific programs are available leading to degrees in technical writing (either in an English department or another department), but they are not offered everywhere and they vary in structure and content. They are, however, the best credential. There seems to be a definite trend at colleges and in English departments to establish technical writing degrees or technical writing concentrations in recognition of the increasing demand for the specialty.

If you are in college and your school has no technical writing major, you can take an English major and add as many technical and computer science courses as you have time for. Most schools offer a few courses in technical writing.

In some parts of the country, colleges offer adult programs in technical writing. One program in the Boston area takes about nine months and offers training in software technical writing. It is not a degree program, but it seems to be well accepted by the business community as a legitimate credential.

You can also get a technical writing job if you have basic writing skills and some technical background or coursework. It can be hard to get the entry-level job, but you would not be handicapped in job advancement if you did not have an English degree or a specific credential in technical writing.

Some kind of technical background or coursework is a requirement. A purely writing background such as in academic and research writing may not help you much.

Given what has been said, this field is open to various backgrounds. As with many other jobs, the difficult part is getting some entry-level experience. You need to develop examples of your work that you can show to interviewers. It is better if you can show them a technical document rather than something else you have written.

## Kinds of Jobs

There is a great variety of jobs in the technical writing field. Some of the major types of documents are:

- Brochures, technical summaries, or handbooks connected with marketing or sales
- Maintenance and repair manuals for computer hardware
- User guides for hardware or software
- Training manuals
- Reference documents
- In-house newsletters and house organs
- In-house standards.

In some ways the writing for each type of document is similar. You must work with a technical person to assure that the information is correct. An editor reviews the document, and the rough draft is changed to reflect the editor's and reviewer's comments. The review cycle may be repeated; then the document is approved and printed.

Some jobs require a lot of running around within the company and talking with people in different locations of the company. That can occur in writing in-house standards, for instance.

Some jobs require a deep technical understanding of the subject area, as with hardware repair manuals and some kinds of software manuals. In these cases, you may spend much time studying the technical aspects, either informally by reading technical books or by taking a company-sponsored class.

You may work mostly with updates of existing documents. It can be difficult to ensure that the update is technically correct and that it is clearly marked in the text. Much technical writing involves that kind of work. You may find it more interesting to write a document from scratch; you may get more of a feeling of its being "your" document and greater satisfaction from completing it. Of course, any new product line or software package requires new documentation. Since the computer industry is continually creating products, there should be opportunities to work on new documents.

There are many writing jobs connected with marketing. The material is not typical advertising copy as for a magazine ad. It is more technical than most ads and involves summaries or descriptions of products that give the buyer enough technical information to know whether the product is what he or she needs. Often these documents represent short brochures or technical summaries that are completed relatively quickly and easily. Writing manuals from the beginning can take months or a year (or longer); you may want to see results more quickly or move on to a new document.

Marketing-oriented jobs tend to be more deadline-conscious, since the sales people need to have the brochures on hand when the product becomes available for marketing. That may require an extra push when the deadline is near.

### Working Conditions and Opportunities

This is office work that is comparatively clean and quiet. Much of your time will be spent working directly with a word processing system or a text-editing package on a computer. A system designed as a word processor is usually easier to use, but text editors on computers have improved greatly in recent years, and many of them are friendly and easy to use.

If you are expected to use a terminal yourself, it is important to have it available in your own office, or at least to know that it will be available when you need it. If several writers share terminals, there should be one half to two thirds as many terminals as writers.

In some cases you may depend on someone else to do your typing or word processing. Here you should be concerned that your work be done quickly and avoid a situation, for instance, where it may be done by a departmental secretary who has many other responsibilities also. In general, the amount of work you can get done is greater if you can work directly at a terminal.

You should realize that in many companies you will be expected to do your writing on a terminal. You should know some kind of word processing or text-editing package or be prepared to learn such a package. The trend is in that direction rather than having technical writers rely on typewriters or pad and pencil.

Problems sometimes arise with the technical people you must work with to review or help with a manual. You may want to try to assess what kind of technical reviewers you will be working with. It is helpful if you can meet them when interviewing for a job or if you can ask a writer currently working with a group whether they have good relationships with their technical reviewers.

Technical writing jobs start in the low salary range but with experience and advancement can rise to the middle and even the higher range. You can advance from technical writer to senior technical writer to principal writer. Principal writer can mean the main or chief writer for a line or set of documentation. There are also opportunities to advance into supervisory and managerial ranks. New supervisory positions and new writing positions are continually opening up because of the growth of the computer industry.

# 14

## Engineering

An engineering job is the most technical of the high-tech jobs. It pays very well, and it also requires more training and education.

To be an engineer, you need good technical and scientific interests. It is not something to pursue if you do not have technical aptitude.

If you already have an engineering degree, you probably don't need this book. You should be able to find a job without too much difficulty, unless the economy is very depressed. What you need to get a job is perseverance.

However, you may be considering going into engineering. An engineering degree requires hard work and time. Even if you already have a bachelor of science degree, you may need two or three more years of work.

Many types of engineers are involved in the computer field in addition to the electrical engineer who designs the circuitry that makes up the heart of the computer. Manufacturing and chemical engineers may be involved in making the product, and a software engineer in designing the software.

In general, as an engineer you would be involved in designing or doing something that has not been done before. You might work alone or as part of a team. You would be given a design assignment, and usually the specifications for the new product are spelled out. With your education and training, you would know how to start the project, and you would know what additional information you needed and have some idea of where to get it.

The design process can be chaotic, especially in the beginning stages. It is basically a creative process; often it involves putting together existing ideas from different places with a few new ideas to come up with a whole that is different and new.

The idea of designing a whole new computer can sound pretty challenging. What happens is that the problem is broken down into small enough segments so that an engineer can handle each one. Obviously the pieces have to fit together, and an engineer or a team of engineers must set forth an overall architecture and design.

Occasionally you will hear of the super engineer who designs a new computer by himself or herself, but the great bulk of engineering work requires not genius but a lot of hard work, sometimes exciting and sometimes detailed and tedious.

### Prerequisites and Training

Although there are occasional exceptions, for an engineering career you need an engineering degree. The work requires a solid technical and scientific background that is not casually or easily obtained. In addition, once the background and degree are obtained, you must learn to apply your knowledge to practical design problems.

Many bachelor's degree engineering programs take five years and require many tough technical courses. If you are in college and considering majors, you should decide on an engineering major quite early because of the amount of coursework required.

You should be aware that engineering departments look for students with demonstrated mathematical and scientific ability and respectable SAT scores.

If you are considering a college, be sure that the engineering program is accredited. You also may want to investigate whether the college has a co-op program (which allows you to spend every other semester working) or some other program to give you practicum experience outside of the college setting. Through such programs you can gain valuable on-the-job experience and build up references that will help you get a good job when you graduate. In some cases, your contacts in a company for which you do co-op work can bring you a job with that company later.

If you are older (for instance in your 20s) and already have a bachelor's degree, you may be considering returning for an engineering degree. Depending on the amount of scientific and technical background that you have, this still may take several additional years of schooling.

You should be cautious about evening courses or adult education courses. For some degree programs, such courses can be applied to degree requirements, but that is often not true for engineering. Most engineering schools require full-time daytime attendance to obtain a degree.

In rare cases you can get into engineering work without a degree. It might occur if a supervisor or manager desperately needed more engineers, and he or she would have to be aware of work that you had done showing that you have good knowledge and ability in a particular area.

Some programs do allow you to earn an engineering degree on a part-time basis, but those are usually offered or cosponsored by a company for employees who are already working in an engineering or a technical job and have a scientific or technical background. Such companies may pay the tuition costs. They may also require or expect an employee to stay with the company for a period of time (such as two years) after completing a degree.

If you are out of college and interested in an engineering degree, write to colleges in your area for catalogues and talk with the admissions people or engineering professors at the schools that have engineering programs. If you are already working for a technical company, watch bulletin boards and company newsletters for announcements of college programs. Your company may also have a central office that deals with college coursework and can give you information and advice about college programs.

### Kinds of Jobs

There is a great variety of engineering jobs even within the computer field.

The electrical engineers are the key people in this field, since they design the electrical circuitry that makes up the computer. The electrical engineer works with schematic diagrams that show the logic of the circuitry to specify the components and the connections between those components. Once that has been done, other people (such as layout designers) translate the diagrams and specifications into integrated circuit chips or printed-circuit boards.

The engineer tests and checks breadboard or prototype constructions of the design to see if it does what is intended. Usually at this stage changes are made in the prototype (and the schematics) to make it work right or work better.

The electrical engineer can work at many different

**105**

levels of the computer, such as the overall architecture or specialized areas such as power supplies. In addition to the central processing unit and central memory of the computer, there are the peripheral parts such as the tape and disk drives, as well as the printers. For these the electrical engineer may design the electrical circuitry that drives them and communicates with the central computer.

The actual design of such mechanical devices may be done by the mechanical engineer. Every part from the smallest pin to the frame and the external covers must be designed and specified exactly before it can be manufactured. All parts must fit together and work together. The mechanical engineer is concerned with mechanical details and with designing and specifying these parts.

The **mechanical engineer** starts with some idea of how much circuitry will go into a computer (how powerful it will be), which in turn indicates how many printed-circuit boards will be needed, which in turn indicates how large a mechanical frame will be needed as well as how many fans or how much cooling apparatus. From these considerations as well as other specifications, the engineer must design a box that is large enough to hold everything and meet all the specifications.

The **software engineer**, who often has a degree in computer science or electrical engineering, designs the software that runs the computer (such as the operating system) or other types of software that are used with the computer. This is a computer programming job, but one requiring more creativity than most. The software engineer often works with assembler-level languages. An assembler-level language may require more work for a specific programming task, although new approaches have been designed to make programming in assembly languages easier. This is one area where a traditional engineering degree may not be needed; that is, it may be possible to work into a software engineering job from one as a computer programmer.

**Manufacturing engineers** become involved when a computer is being built. They determine what manufacturing equipment is needed or how existing equipment must be modified. They must continue to monitor the manufacturing process to see that quality is maintained and production is efficient. The manufacturing engineer may also become involved in modifying or even building new pieces of equipment or developing new ways of using existing equipment.

**Chemical engineers** can also be involved in the manufacturing process. Some processes, such as printed-wiring board manufacturing, involve many chemical solutions that must be specified and monitored. Slight contaminations or problems in the pH (level of acidity) of the solutions can ruin the boards.

**Acoustical engineers** may become involved, for instance, to muffle a noisy printer that must be used in an office environment. Engineers who work in quality control are called **quality engineers**, and still other engineers may be concerned with product safety.

### Working Conditions and Opportunities

Much engineering work is office work, involving designing something on paper or on a computer-aided design (CAD) system. However, testing something, such as a prototype printed-circuit board, may require working in a lab with electronic equipment.

Manufacturing engineers may work in the manufacturing facility itself and with the manufacturing equipment. But even so they would have an office.

Work may often need to be done on a deadline, especially in the competitive computer industry. If a new design is slow to get to market, a competitor may come out first with a similar product, and your company would lose sales and money.

If you are looking for an engineering job, you may want to check to see that you would have the necessary resources. For instance, for most jobs you should be using a CAD workstation in your design work. This system can speed the design process and do tedious chores such as redrawing a schematic diagram quickly after it has been changed. A CAD workstation can even test the design with computer simulation so that when the prototype is made it is more likely to work correctly. Other resources might include lab facilities or access to parts and materials. If you are a new engineer, it would be good to identify senior engineers who can give you help or advice if a problem arises.

Engineers, especially electrical engineers, are well paid. That and the fact that the work can be creative and interesting makes the field very attractive.

Engineers can move on to be project managers and move up into management. Some companies have what is called a professional consulting engineering track that allows an engineer to advance to a high level in salary and status without going into management.

Compared to other workers in the computer industry, engineers often have the best opportunity to start their own company. This can be as a consultant offering services to other companies, or manufacturing a product that you have designed. The major companies dominate the market for general-purpose computers and are difficult to compete with. However, there are openings for specialized computers or specialized electronic parts or devices.

# PART III

# 15

## Pumping Up Your Credentials

It is very important to get the most out of the credentials and background that you have. Most people have some kind of background that may relate to a high-tech job—some familiarity with electronics or computers such as being able to use a computer terminal.

If you are looking for a job, especially an entry-level job, you need all the help you can get. You need to spell out that kind of experience in your résumé and use it in interviews. This should not be done in a heavy-handed way, but simply to call attention to something in your background that may be relevant.

Although you might believe that you have no useful background, often experiences you have had that may not sound relevant to a job could prove to be quite helpful.

Always look at your background with your goals in mind. It is better to find experience that relates to the job you want to get than just to list a variety of experiences that are not closely related.

However, if your résumé is weak technically, you may want to list some experiences that give it a technical "flavor."

Experiences that may be important personally to you, such as poetry you have had published or meditation retreats you have gone on, often carry no weight with interviewers. You have to find "technical" experiences or use your imagination to relate experiences to technical areas.

### Looking at Your Background

There are several steps to looking at your background and using it more effectively.

Look at your background carefully for any technical experience you may have had. This can include playing around with a friend's home computer or any kind of computer course you may have taken.

Relate this experience to the high-tech industry. For instance, if you had to look at an electrical diagram of a car to repair a broken tail light or horn, you might say in an interview or a résumé something like: "I had to understand the electrical logic of the circuit schematic in order to make this repair." Do not be pushy or obvious. Try to give an impression of familiarity with technical material. Do not claim things that you cannot back up, or claim to have expert knowledge. You don't need to understand the electrical circuitry of a car inside and out to be able to say: "I saw that the ground wire for the horn was broken and knew that was the reason the horn did not work." The main goal of "pumping up your credentials" should be to get the most out of experience that you have had, not making up experience or saying you can do things that you cannot. The latter can only lead to embarrassment if the interviewer pursues a line of questioning to see what you really do know. It can lead to trouble if you get a job but do not know enough to handle some of the responsibilities that you claimed to be able to handle. Sometimes you can rescue yourself in such situations by some quick studying or by relying on coworkers for a while, but that is risky. In a job situation, unless it is clearly labeled a training position, you will be expected to step in and handle your tasks quickly.

You can exaggerate to some extent and also spend some time thinking of the best way to phrase things. For instance, in the example given above, you might have said: "I fixed a car horn by replacing a broken wire." It is more effective to talk about understanding the "logic of the electrical circuits," or being able to "read the circuit schematic."

I must also point out that you will encounter interviewers who tend to downgrade this kind of experience or give no positive reaction about it. They are interested only in more formal credentials such as courses or on-the-job experience. Don't worry too much about those interviewers; you may get a positive reaction from them only if you have a degree in computer science. Move on to the next interview. Remember that, even if it is not apparent, you may still be giving a better impression than you would otherwise, and you may be surprised by a request for a return interview or a job offer.

### Experience to Consider

Many kinds of experience can be related to high-tech jobs. Some that should be considered are the following:

- Use of a computer terminal. Any kind of hands-on experience with a terminal is valuable. This can include word processing terminals, terminals used in retail stores (for instance, for inventory purposes), and small home computers. Be as specific as you can about the model of the computer and the terminal; if your experience is not very extensive, you can use such phrases as "familiar with the DEC VT220 terminal" (even if you only punched a few keys), or "used Apple

113

Macintosh for word processing and typesetting" (even if it was only one letter on a friend's computer).

- Computer demonstration. This includes any kind of demonstration that you have viewed and that has given you some understanding of the capabilities of the computer or terminal. Retail stores selling home computers often put on demonstrations. Again, such phrases as "familiarity with" and "understanding of" are useful. Avoid saying that you can operate a computer with which you have not actually had some hands-on experience. Sometimes you can gain that experience during demonstrations if people from the audience are allowed to try the computer.

- Computer programming. Any kind of programming that you have done is valuable, even a very simple program on a home computer. You should be able to say exactly what kind of computer it was, what programming language was used, and describe the program you wrote. You can also list or describe any computer software packages you have used. A software package is usually designed for a specific purpose, has fewer commands, and is easier to use (although more limited) than a programming language. Again, be specific about the package: statistical, accounting, or word processing.

- Electromechanical repair. If you have ever done any kind of repair work on machinery that has electrical circuits, you may want to list that experience—including weekend work on the family car. If you ever have had to track down and figure out an electrical problem, you may be able to say that you had to understand the electrical circuits, or if you looked at the circuit diagram, that you solved the circuit schematic.

- Electronic kits. If you ever assembled an electronic device from a build-it yourself package, that is good experience to list and describe. These kits point to skills others may not have developed.
- Soldering. Describe or discuss any soldering that you have done, whether in an electronic kit, car repair, or simple home electrical repair.
- Use of electronic meters. This includes any of the meters for voltage, resistance (ohms), or amperage, or an oscilloscope.

Any of these experiences can beef up an undernourished résumé and give your background more of a technical feeling than it might otherwise have. By themselves, they may not get you a job, but combined with other credentials or coursework, they may do the trick.

# 16

## Coursework

Coursework is the most traditional and time-honored way to build up your background, acquire valuable knowledge and skills, and get a good job. That is certainly true in the high-tech area, where almost everyone must have some kind of technical knowledge and background.

Even so, before engaging in a lot of coursework you must consider what you need; that is, pursue coursework when you have thought out your goals and it is clear that you really need it.

Often you can get a lot of mileage out of your present background and pumping up that background appropriately. By updating your résumé and sending out a few exploratory mailings, you may find that your present background is adequate. If you get some nibbles, you may want to proceed directly to the job-hunting stage or consider taking a course while you are job-hunting.

As has been said before, if you can get your foot in the door and start with a company, many companies offer in-house training programs or may be willing to pay tuition for college courses.

It is clear, however, that if you do not have a high school diploma you should definitely obtain one. There may be some situations in which you can get a high-tech job without it, but they are rare. Find a program for adults who want to get a high school diploma part time or in the evenings.

### Set a Learning Goal

It is important to have a clear goal for any courses that you take. Your goal may be simply to explore a few high-tech areas; even so, you should have some feeling for what your interests are and what possibilities are realistic. Otherwise you risk spending a lot of time and ending up with little in the way of a coherent or focused background that might appeal to an employer.

The most valuable coursework is that which aims at a particular job, such as programming coursework that aims at a programming job.

You should have done some work on exploring different possibilities and your interests and focusing on one or two jobs that appeal to you. It becomes easy to set up a plan of coursework when you have a clear idea of what you want to do. In computer programming, for instance, your choice of coursework is obvious. Since many schools offer courses, it becomes a matter of choosing a program that fits your budget, that interests you (scientific as opposed to business programming, for instance), and that fits other considerations such as your time schedule or proximity to where you live.

Your coursework and job goals must be realistic and in line with the time and energy you can afford. If your goal is to become an engineer, you must settle down for a long haul of full-time schooling and find a way to pay for it. Some people have pursued this goal by taking out a second mortgage on their home; that is clearly a major decision that must be considered carefully. However, if you see no way to finance years of schooling or have little desire to be a student again for such a long time, your goal must be reexamined.

For instance, a year or so of computer programming courses may establish a good background in that area, and you may find an opportunity to work your way into a job as a software engineer (or be able to find a company that will send you back to school for training as an engineer).

Like most people, however, you will probably be considering two or three courses taken at night school. Since it can be a strain to take courses and work full time, you need to choose wisely. Feeling that the courses are leading in a definite direction can help to keep you motivated and ensure that you to get the most out of each course.

### Types of Coursework
There are several basic approaches to coursework:

- Full-time coursework at an accredited college leading to an associate's or a bachelor's degree.
- Full-time coursework at a technical school, such as a nine-month program in electronics or computer operation.
- Evening coursework at an accredited college leading to a degree.
- Evening coursework aimed at building up a background of several courses in a specific area.
- Correspondence coursework such as a course in electronics.
- Part-time, company-sponsored coursework at an accredited college.
- In-house training program at a company that you work for.

The choice among these alternatives will depend on your goals. Courses at an accredited college will give you the best background and the best credentials. The next best thing is to complete a structured program leading to a degree or certificate: a college degree program, or a nine-month program at a technical school.

Of course, if you have access to a company-sponsored training program or coursework, you will probably want to pursue that approach. If you change jobs within the company, interviewers will be familiar with the program and will view it as a legitimate background.

A correspondence course (through the mail) may not always be looked on as valuable, yet you can learn a great deal and it would be better than nothing. You may want to pursue this approach if your schedule interferes with other kinds of coursework or if you have a personal preference for doing your schoolwork at home. In a correspondence course you must exert greater self-discipline because your studying can be set aside by other demands. You also forfeit the ability to talk over problems with a live teacher.

Selecting a school is also important. If you have lived long in an area, you are aware of colleges near you, and you can ask the admissions office to send you brochures and catalogues. You may want to contact large universities that are at some distance from you; some of them have extension night-school programs that are located off of the main campus. Some large urban universities have such programs in the suburbs, for instance, in local high schools in the evening.

If you plan to take college courses, find out if the college is accredited. If you are pursuing an engineering degree, you want an engineering department that is accredited (this is separate from the accreditation of the college as a whole). Some technical courses offered as part of a night-school program may not apply to an engineering degree. These matters can best be determined by contacting the engineering department.

Technical schools that offer programs in computer operation, electronics, or programming are usually not accredited. They do not grant degrees, and they offer programs with a duration of, for instance, nine months. Do not confuse these schools with technically oriented colleges that grant two- or four-year degrees. If you are in doubt, ask the admissions people or read the literature carefully.

These technical schools often advertise on TV or in the newspapers and may be listed in the Yellow Pages.

In choosing any college or school, it is usually better to go with a well-established one. The advice and comments of friends or other contacts are invaluable. That is by far the best way to get an idea of what classes are like at a particular school, the quality of the instructors, and whether computer terminals and hands-on experience are readily available.

Do not assume that because a college is a full-fledged degree-granting institution with a good general reputation, its technical or night-school courses will be as good as they could be or meet your particular needs. Ask around to find out its reputation for its technical and scientific departments.

# *17*

# *Résumé Writing*

The résumé is extremely important in your job-hunting efforts. Often it is your first contact with a potential interviewer, and often it is your *only* contact, since many interviewers answer phone calls with instructions to send in your résumé. This can be frustrating when you have rehearsed what you might say and are bursting to tell an interviewer about your qualifications and eagerness for a particular job.

Your résumé must make a good impression and be fairly complete. It should not be too wordy, nor too short. If possible, it should be written for the specific job you are seeking.

If writing a résumé seems intimidating or you have never written one before, you may want to use one of the résumé writing services, but I would recommend taking a crack at it yourself first. Several books are available that give examples of résumés and ideas for writing them. Some of them are listed in the reading list.

If you plan to use a job agency or to see a job counselor, you should be able to get advice on how good your résumé is and how it might be improved.

If typing a neat résumé is a problem, consider writing it yourself and finding a typist to type it for you. For a few dollars you can have a professional-looking résumé.

I cannot emphasize too strongly that the résumé must *look* well. There must be no typographical errors, and it must be neat. Some people go to the trouble of having their résumé set in type and printed. I do not recommend that unless you know of a print shop that provides the service at a flat rate.

Another argument against printing is the importance of keeping your résumé up-to-date and making changes in light of advice about it or new information that you receive. Also, you can work up two or three versions of the résumé slanted toward various kinds of companies or different jobs (which should be closely related).

The whole business of résumé writing can seem time-consuming, especially if you are doing the typing yourself. But it is time well spent. I cannot emphasize that too strongly. Your résumé is your first impression on an interviewer; its neatness and relevance to a particular job will go a long way toward establishing you as a likely candidate.

### Elements of the Résumé

A résumé is divided into several major sections. Items within each section should be listed in reverse chronological order; for example, the experience section of a résumé might look something like this:

---

EXPERIENCE

---

| | |
|---|---|
| 1991–Present | Network Administrator, Superduper Computer Services. Maintained a network of 20 Model 727 personal computers, two Supertech 60 computers, three ZY-2 disk drives, two LPXX line printers, and one QLP laser printer. |
| 1988–1991 | System Administrator, Supertech Computing Corporation. Administered a Supertech 50 computer with two XY-2 tape drives, three ZY disk drives, LPXX line printer, and PP-5 letter quality printer. |

1986–1988    Computer Operator, Allied Computer Services. Second-shift operator, Supertech 30 computer with one XY-1 tape drive, one ZY disk drive, and one LPXX line printer.

This example shows important points. It lists the date, the type of job, and the name of the company. In addition, it lists the equipment that the résumé writer used, which is very important. An interviewer will want to know what kinds of equipment you are familiar with.

In addition, a technical approach or turn of mind is precise rather than vague, and being able to list in detail the equipment you have used makes a much better impression. "Used a Wordamagic word processing system with three DD-500 disk drives" is much better than "used a word processor." In this example the interviewer would also understand that you are familiar with a sophisticated processing *system* that can support several users rather than a simpler processor designed to be used by only one person.

The major categories of the résumé are:

- Personal information. Usually placed at the beginning of the résumé, this includes your name, address, and phone number. The trend is away from including such information as weight, height, age, marital status, and health status. If that information is desired, the interviewer will ask for it.
- Objective. This second category, which is optional, states the kind of job you want, such as "technical writer," or "field service technician." Here you can indicate clearly that you are looking for a promotion beyond your present job title, or you can state your job goal. If you use this category, you may have to prepare several versions of your résumé with different statements.

- Skills. This optional category can comprise a few sentences, describing particular skills that you can bring to a job. Again, it may be slanted in one direction or another depending on the job you seek. You should not be modest in this section (but do not exaggerate too much).
- Education. This section, listed in reverse chronological order, may be placed before or after the experience section. If you are just out of school, usually the education experience is listed first.
- Experience. This section lists your job experience in reverse chronological order. Emphasize and describe in more detail jobs that are technical or relate to the job that you seek. You may also list other kinds of technical experience in this section.
- Interests. This is an optional section and should be short. Even though these experiences may be important to you and make your life richer and fuller, it is best to omit them or list them very briefly unless they relate in some way to the job you are seeking.
- References. These refer to people who can attest to your good character and personality. Most résumés say "Supplied on request." You should be prepared to offer references of people working in the field, executives, ministers, and so on. However, interviewers usually are much more interested in the names of previous employers or supervisors who know you and your work.

Interviewers usually want to be able to trace your job history from the time you graduated from college or high school. The experience section usually serves this purpose. It is best to be able to show a continuous experience of work or fruitful activity. If for instance you spent six months hitchhiking around the country after you graduated or were unemployed for a period of time, you can be honest about it, but you may not want to list it on your résumé.

Some people choose to list only experience that is the best or most closely related to the present job application. Be prepared, however, to give a chronological listing of what you have been doing with your life. If you seem to be covering up a period of time, interviewers may wonder whether you are trying to hide a bad job experience.

### Style and Strategy

Your general approach is to be as positive as possible and to present your experience in the best light. Any job experience or courses related to the job should be described in detail. Avoid dwelling on other matters even though they may be important to you personally.

Some mild exaggeration is okay. Everyone tends to do that on résumés, and you will put yourself at a disadvantage compared to other people if you do not do so.

However, do not claim skills and knowledge that you cannot back up. Remember that either in an interview or on the job you may be asked to demonstrate your skills or knowledge. If you have made a false claim, it can only be embarrassing. For the same reason, avoid claiming degrees that you do not have. Such statements tend to backfire, and you will regret them sooner or later.

There may be times when you want to omit something from your résumé. This is a tricky question that you must decide for yourself, since it is always safer and easier to be honest. For instance, you may want to omit a brief period of unemployment by stretching the period that you worked during the previous and following jobs. You may also want to omit an advanced degree such as a master's in French literature, because you may be considered overqualified or not really interested in a technical job.

The "overqualified" label can be one of the most annoying and unnecessary. It can be especially frustrating to people whose degrees do not give them much in the way of job skills and who need to work and develop those job skills like everyone else.

One school of thought is to say as little as possible in the résumé; say just enough to get a job. List just enough credentials and background to indicate some qualification for the job and to tantalize the interviewer. It is true that the longer the résumé, the more possibility you have of saying something that will rub the interviewer the wrong way. A résumé of one page is best, and two pages should be the maximum.

My own feeling is that a short résumé may be best if your credentials are weak; however, you should list any good technical courses or background that you have.

# *18*

# *Job-Hunting*

Now is the time to go out and start looking for the job you want. You should have your goals clear by now, you should have the credentials you need, and you should have worked up your résumé so that interviewers will respond to it.

The unpleasant experiences that people have when job-hunting often arise because they do not have clear goals or are applying for jobs way over their heads. They have an unpleasant interview experience or become discouraged at not even being able to get interviews.

Job-hunting can be exciting and interesting if you have the right attitude. You should not pin all your hopes on any one interview or any one company. You need to be as relaxed as you can be. You will be meeting a variety of people and learning a lot about companies and job situations. Even if you do not get a job with a particular company, you should be filing away what you have learned about it; you may be interested in it again in some future job situation.

The most important thing is to maintain a good attitude and to keep plugging away. That can be difficult, but if you keep trying you will succeed. In fact, the secret of success is simply perseverance. Too many people quit too soon or take a job that is not really satisfactory. You deserve to get a good job.

### Job-Hunting Goals

By the time you reach this stage of the job-hunting process, you should know what jobs you are looking for. You should have some idea of the salary range. Your background should at least make you a possible candidate for the job.

There are other more specific goals that you should set either before you start job-hunting or soon after. These involve more detailed concerns and requirements that you may have for a job.

- **Commuting**. You need to set a limit as to how far you are willing to commute. This can be based on mileage, though more often people think of commuting time. Many people do commute for as much as an hour each way, but most people want a shorter travel time.
- **Salary**. You should try to learn the salary range for the job you are applying for; set a minimum for yourself that is also realistic for the job. You may have to consider your expenses and what you can live on and decide whether you can take a lower-paying job that will, however, establish you in a good field (and set you up for future financial rewards). Fortunately, the salary levels for high-tech jobs (even at entry level) are very attractive for people who have been working in other areas.
- **Contract work**. This temporary full-time work usually pays better than permanent full-time work because you receive no fringe benefits. The trade-offs are security versus money and taking long vacations. Permanent positions are better if you are ambitious about advancing rapidly. In

some cases, you can get a permanent position after showing your worth in a temporary position; but don't count on that unless it is indicated clearly. The main danger of contract work is that it is temporary and you are viewed as expendable (in times of budget crisis, or if a project is canceled or refocused).

· **Defense work.** Some people prefer not to work for companies that work on military projects, or at least not to work on military projects themselves. You need to think this through carefully so that if a job comes along that involves a military project, you know where you stand. Interviewers (at a job agency, for instance) may encourage you to be open to all kinds of jobs but usually will respect your decision. Their main concern is to avoid sending you on an interview for a job that you would not accept. Many companies do military or government work at one time or another; it may require adherence to military specifications, involve annoying security procedures, and be very deadline-oriented.

· **Interview questions.** Now is the time to consider (and even write down for yourself) other specific questions that you want to deal with in an interview. You may want to ask about company benefits, or find out if you will have good equipment to work with (for instance, if you are a layout designer, or a word processor). You may want to ask about the general salary range for the job and advancement and career paths. Interviewers expect some questions, and it is right to ask them; but be careful about asking about a specific salary (that is tricky and is discussed later in this chapter), and avoid dominating the interview.

Try to get as clear an idea as you can about these goals. Write them down or write about them in your journal. Discuss them with your spouse or friends.

Remember that you may hear about or get an interview for a job that does not meet your goals (such as one that is too far away). You may be sorely tempted to reevaluate your goals, and you may need to do so if the opportunity is especially good. But also remember that opportunities that sound good may not be so in reality, and another may soon come along that does meet your goals.

### Making Contact

Making contact with a potential employer and getting an interview can be the hardest part of getting a job. You have to get the interview before you can sell yourself to the interviewer. That is why the résumé can be so important; as has been said before, it is often your main preliminary contact, and it may be your only contact if you are looking for an entry-level job.

Before sending in a résumé, you must find out whether a company has any appropriate job openings. You need to find out the right place to send a résumé. Since a large company may have several personnel offices and different individuals may handle applicants for different departments, you need to find out the right person to whom to send the résumé.

Some companies encourage phone calls, and some request only that a résumé be sent. In most cases, however, the interviewer will request a résumé before setting up an interview. If, after sending it in, you get no response, it is usually wise to follow up with a phone call after a week or two. If the person you speak to says they are reviewing résumés and will get in touch with you, don't be afraid to follow up with another phone call. By calling back you show that you are interested in the job.

It is usually best to speak directly with the technical manager of the department that has the job opening rather than personnel people. The problem with going through personnel is that they may be flooded with résumés and all begin to look alike, unless one is truly outstanding. If you can speak directly to a manager and sell yourself to him on the basis of your background or enthusiasm or interest in the job, you have a better chance of getting an interview.

A résumé that you send should have a brief cover letter. You need not go into a lengthy discussion of your background, though you may mention one or two items that are particularly relevant to the job you are applying for. Indicate clearly what job you are applying for and ask for an interview.

Leads for companies to send résumés to can be found in several ways:

- **Classified ads.** This is the traditional way to get an interview. Determine which paper carries the best ads in your area. Usually your best bet is the classified section of the Sunday edition of a large metropolitan newspaper. Be prompt in sending out your résumé or phoning (if the ad gives a phone number).
- **Job agencies.** This refers to agencies that are paid by the company, not by you. Many companies use agencies rather than classified ads so that candidates are prescreened. Such an agency will in fact want to interview you first and may provide helpful advice such as changes to make in your résumé. Be open to this advice, since the agencies are in close touch with the current state

**131**

of the job market. If you are looking for an entry-level job, the agency may not want to handle you. Don't be discouraged by that. You may still be able to get good advice either through a phone call or an interview. Ask friends about agencies that they have used, or look in the Yellow Pages.

- **Job fairs.** Job fairs are occasionally run in high-tech areas. They are usually advertised in the papers and aimed at recruiting experienced technical professionals. They may have representatives from many companies and be run either by a government agency or a private company. The formats vary; you may have to register and be interviewed by someone from the sponsoring company or agency, or you may be able to walk around from booth to booth and talk with personnel representatives directly. You should have several copies of your résumé with you and be prepared to have a brief interview on the spot; you will almost surely have to have another interview at the company before a job offer is made. You may have a hard time if you are looking for an entry-level job, but it is worth a try. You may be able to get brief interviews that would not have been possible if your résumé were sent in cold to the personnel department or in response to an ad. The job fair can also be a unique opportunity to find out what companies are looking for and what kinds of jobs are available in your area.

- **Alumni.** You can sometimes use contacts among alumni of your high school or college. In most cases an alumnus will be glad to talk with you. He or she need not be a manager of a department that has job openings. Usually you can get inside information about what it is like to work in a particular company, and perhaps names of managers or other contacts in departments who are looking for job applicants. Your placement office may have lists of alumni working at various companies, or you may contact old friends to see where they are working or if they know other alumni working in high-tech companies.

- **Friends and other contacts.** Again we come back to some of your most valuable contacts: friends and other people you know. They can help in much the same way as alumni. Of course, the most valuable contact is a friend or personal contact whose department has a job opening. He or she can give the manager a personal recommendation of your abilities and deliver your résumé directly. This kind of introduction is invaluable, because it sorts you out from a crowd and may predispose the manager to hire you even before your interview.

Counselors or psychologists talk about the value of "networking." A network is simply all the people you know (including friends and acquaintances). The more people you know, the more likely it is that you will have someone who can help you get a high-tech job. If you don't know many people (for instance, if you have just moved), you may want to try to build up your network. That means introducing yourself and getting to know people who may be able to help you. It can be done at church, at parties, in a bar, or at almost any kind of social gathering.

Your goal should be simply to meet people and establish some kind of contact with them. Think of it as building up your resources and potential sources of help, rather than pressuring people immediately for advice and help.

You should not feel cynical or manipulative about this. You are constantly meeting people and expanding your social network whether you think about it or not. Most people are glad to meet new people and to give advice or other help. People like to feel helpful and recognize that someone else may be able to return a favor in the future.

A strategy that is particularly difficult is the completely "cold" call. By this I mean phoning or sending résumés to personnel departments or technical managers without knowing whether they have any job openings. This can be very discouraging, and it requires great perseverance.

If you have no contacts or leads, you may feel that you have to take this approach. Be prepared for a limited response to your inquiries. In many ways it is like phone sales: Many calls must be made before a sale is made. Take that attitude, and don't be discouraged by the occasional rude or unsympathetic response.

Above all else, and this applies to all phases of job-hunting, do not blame yourself or get down on yourself for slow progress. Remember that the going may be slow at first, especially if you are looking for an entry-level job or if the economy is depressed. For every success story you hear, there are ten stories (if not a hundred) of slow progress before a job was found.

### Interviews

Once you get the interview, you have to go to it. The interview is the real gate to success or failure. The interviewer will be deciding whether you are really interested in the job, have an agreeable enough personality to work well with others, show basic enthusiasm, and whether the job fits your goals as well. The interviewer wants to feel that you will fit in with the company and be happy in the job (so that you don't quit in a few months).

You must take the interview seriously and yet at the same time not be too nervous or try too hard to impress. It is okay to be a little nervous; interviewers expect that.

You must also keep in mind what kind of job you are interviewing for. If it is a sales job, you need to be much more enthusiastic and outgoing than for, say, a programmer's job. The interviewer for the programmer's job does not expect a bright and outgoing personality, but more of a technical personality, quiet and introverted.

You do need to be somewhat outgoing, and that is not a problem for many people. It is a good idea to ask questions about the company and the department you will be working for. Interviewers expect it, and it is an opportunity to find out things that you want to know.

It is also a good idea to learn something about the company and its products before you go to the interview. You may be able to obtain sales brochures from a sales office or a copy of the annual report from the financial office or corporate headquarters. Interviewers like to feel that you are interested in the company beyond applying for a specific job. If you have a lot of interviews, this can become difficult and time-consuming. Sometimes if you arrive early for an interview (particularly in a personnel office), brochures about the company may be lying around that you can read. As always, you should rely on friends and contacts who have worked at the company or know something about it. If you are engaged in a long job-hunt, you can keep your eye out for articles about various companies in your area in the business section of the newspaper. It is also a good idea to read some magazines on high-tech companies, such as *Computerworld*. The advertisements can tell you a great deal about the companies.

The mechanics of interviewing are as follows: In most cases you will be telephoned after sending in a résumé, usually by someone from personnel. You will need to set a time for an interview. You may be asked to set aside most of a morning or afternoon; however, the length of the interview depends on the job and the company. You should ask how long the interview will be and whether you will be interviewing more than one person. In some situations you may spend as little as half an hour with one person. In others you may have a longer interview with a personnel person, then one with a technical manager, and perhaps be introduced to future coworkers and shown the office or working space that will be yours.

If certain times are bad for you, be sure to say so. Don't take an interview at a bad time, or one that will require that you rush from one thing to another (you will be too nervous at the interview). Interviewers will want you to come in during business hours in most cases, though sometimes an interview can be scheduled very early in the day or at the end of the day if you have other conflicts (such as a current job).

You may be asked back for a second interview. That usually means that you are at least considered a likely candidate, but it may only mean that an important person, such as a manager you would be working for, was out of town during the first interview.

It is okay to ask at the end of an interview how you are being considered. Most often you will be told that they are conducting other interviews and will be in touch with you in a week or two. Feel free to call back if you don't hear from them; that shows that you are still interested in the job.

If they say they don't consider you suited for the job, it is best for you to know it rather than build up your hopes and have them fall later. It also gives you an opportunity to ask what may be lacking in your background. It can be painful to hear negative feedback, but it can also be valuable; if you keep hearing the same answers over several interviews, you may realize that you are applying for a job over your head or that you really need to consider taking appropriate coursework.

Perseverance can gain success. The fact that you are gaining interviews shows that you must have some good points in your résumé, and eventually someone will make you a job offer.

Of course, there is always the possibility that you have some problem in your interviewing technique that you are not aware of. Sometimes a job counselor or your friends can give you feedback on this. Few of us have such refined interview skills that they could not be improved.

You should dress comfortably and conservatively for the interview. In most cases, men should wear a tie; women should dress conservatively. Both should avoid loud, flashy, or unfashionable clothes. Avoid getting labeled as offbeat.

Try to be relaxed at the interview, but alert. You may be asked to explain or defend any questionable or unclear parts of your résumé. If your background is weak, you may be asked why you think you can do the job. You should be interested in the company and the department and not be too silent. But don't be too talkative or try to dominate the conversation; remember who is doing the interviewing. Be pleasant and conversational; avoid being argumentative or making negative statements about the job or the interviewer. If you have reservations about some aspect of the job situation, keep them to yourself as much as possible; ask questions tactfully as though you were merely interested in the matter.

If you feel that you truly could not take a job if it were offered to you, say so at the end of the interview.

Even if the interview does not go well, remember that it is an opportunity to practice your interview skills and to find out more about a technical field or a particular company. You may also be able to get important feedback about your background and how you present yourself in an interview. Feedback can be painful, but always evaluate it carefully and take it with a grain of salt.

The interview is also a good place to get an idea of the basic aspects of the job, inluding the salary or wage range. It is usually best to avoid being too specific about money at this stage. You may want to inquire about daily schedules, time off for lunch, and benefits and to get an idea of future opportunities and career paths in the company. You may also want to find out what kinds of equipment you would be using, such as a word processing terminal.

Role-playing is a valuable way to get practice in interview situations and some feedback on how you function. It is a technique from psychology or counseling that is not difficult. If you are seeing a counselor, he or she may be able to help you; if you have friends who are counselors or actors, they too may help.

The basic technique is to pretend that you are in an interview situation. Get a friend to be the "interviewer" and to have in mind a list of questions covering such topics such as your job history, your interest in this job, your background, your goals for the future, and so on. You and he or she should go through the whole interview process, beginning with entering the room and saying hello. Be sure that the interviewer includes some questions that you are nervous about, such as "You don't seem to have much technical background; do you think you can handle this job?" The interviewer should not be too difficult but try to be realistic. Most people who have not tried such a technique before find that five minutes or so is enough for a first session.

The technique sounds artificial, and it is. But it can be valuable in giving you a chance to rehearse your answers to difficult questions and making you feel better prepared for the real interview. You can also have a silent third person as an observer. His or her feedback and that of the interviewer can be very helpful.

### Getting the Job

Now is the time when your hard work pays off and you have received a job offer or a company has indicated that they will be making an offer soon. Now is the time to take care of any last-minute issues, to make sure that the salary is what you expected or need, or to make sure for yourself that the job is really what you want.

Waiting to get a job offer, even if you have had a good interview, can test your patience. From the time that you first put in your résumé to the time that you accept a job offer can take two months or longer. You will have spent this time going to interviews and being checked out by the personnel office. The procedure can vary, but the personnel office will usually want to check out your résumé and references (including previous supervisors and managers) after you have had a successful interview with a personnel representative and a technical manager.

The job offer will specify a salary or wage. For some jobs this is a fixed rate, and the amount will be known when you apply for the job or early in the interview process. More often there is some gamesmanship and subtle negotiating involved. You may be asked during your interviews what your minimum is. Be very careful about your answer. Some people advise being cagey and proposing that the company make an offer. This is

**139**

based on the theory that the company may make a higher offer in order to ensure that you will take the job. It can also work against you if the offer is too low, although you can then make a counterproposal for a higher amount of money.

If you do state a minimum, you will be taken at your word and will probably receive an offer at that level. So if you do indicate a salary level, be sure it is one that you will feel comfortable in accepting. If it is too low, you may end up feeling that you have been shortchanged.

Some points are important in negotiating salaries and getting the most credit for your background. An advanced degree such as a master's or doctorate should be worth something in salary even if it is in a non-technical area. Don't be put off if an interviewer implies that the degree is not relevant to salary. Most companies recognize the value of such a degree and make a higher offer. This is appropriate since the degree indicates that you have a higher level of intelligence and additional abilities, some of which should be transferable in some way to your new job.

Before you accept the job is the time to deal with other issues of job conditions. For instance, if most people in the department have computer terminals in their offices, you may want an assurance that you will have one in your office.

You may also want to clear up last-minute questions about benefits, the exact location where you will be working (if this is in doubt), whether your department may be moving to a new location (as occurs occasionally in large companies), whether you will have to travel, whether there is any possibility that you may be "loaned" out to another department, opportunities for advancement, and any other questions that come to mind. If these issues are relevant to the particular job, they will probably have been discussed in your interviews, but you may want to have them clarified.

Most of all, you need to consider the job offer carefully and feel sure that it meets your needs. If you are desperate for money or exhausted by job-hunting, you may want to take the first job offer that comes along. That is all right, but always try to give yourself a little time to consider the situation. Even if the job is not quite what you want, remember that as little as one year's experience will move you out of the entry-level category; then finding another job that is really to your liking should be much easier.

In many cases the job will be what you want, especially if you have devoted some time to the exploration and preparation stages of job-hunting. You can look forward to it and know that you will be building up valuable credentials for the future.

# 19

## *Your Mental Outlook and Attitude*

The real key to success is perseverance and maintaining a positive attitude. You must believe in yourself and in your ultimate success in spite of setbacks and slow progress. You must have a positive image of yourself and a positive image of yourself traveling toward success.

You may have heard all this before, and it is often simple to say, difficult to carry out. There are numerous courses, programs, and weekend seminars that aim to accomplish this kind of self-image boosting. These programs deal with self-confidence, positive thinking, positive attitude, self-empowerment, self-actualization, self-suggestion, visualization of positive imagery, positive energy, and so on.

If you are interested in this kind of program, you should determine whether a particular program appeals to you and fits you individually, whether it fits your budget and time schedule, whether you know people who have taken the program, and whether you may be able to take it together with a friend.

Remember that any form of counseling, individual psychotherapy, or group therapy can accomplish much the same goals. However, it may take time before you begin to feel results and to feel more energy and confidence in your daily life.

The positive-thinking kinds of programs often have a more immediate effect (and at the same time do less to change or deal with underlying personal problems that are a long-term drain on your energies).

That brings up a key point: the question of where you can put your mental and emotional energies. Worry, discouragement, poor personal relationships, lack of confidence, personal problems, unresolved psychological issues all create a great drain upon our psychological energies. The extent to which you can deal with these problems is the extent to which you can free up more energies, which in turn translate into activity and action that lead to growth, better relationships, and better living and working situations.

The positive-thinking programs deal directly with our self-confidence and insecurities. They try to give us positive images and feelings about ourselves that in turn make us feel more energetic. Each of us also tends to behave in accord with our image of ourselves and an image of the future that is tied to our expectations. A positive image tends to generate positive action that fulfills that image; that is discussed later in this chapter.

Next is the matter of being realistic in our goals. That seems contradictory to what has been said about positive thinking, for "realistic" sounds like a downer. Reality usually is a downer; our fantasies and images are often ahead of what is available in reality.

But I believe strongly that the best success comes from a positive attitude and positive thinking that are on one hand well grounded in reality and the possible and on the other hand based on a vision and positive image of ourselves that knows each of us can do great things and achieve a satisfying success.

In relation to job-hunting, the greatest problem people have is being too realistic or too optimistic. If you wanted a job as an engineer designing computer chips and yet had only a high school education, you would be bound to be disappointed. Yet if you think that you have no qualifications or abilities and do not apply for good jobs, but take the first offer you get, you are again bound to be disappointed.

It is the balance between reality and ambition that leads to success. Ambition in a vacuum becomes pleasant daydreaming. Reality uninspired by vision becomes dull and oppressive.

I must give a warning about popular psychological programs. If you have not dealt with much psychology or therapy, many of these programs will be exciting and stimulating when you first finish a weekend. That is because new perspectives have opened up on yourself and your world, and your psychological energies are loosened up. That is valuable and to your benefit.

The danger, however, is in being bowled over by the experience and becoming tied in with a particular organization, in effect becoming a convert and a fanatic about it.

Remember that the excitement and new perspective can be obtained in a wide variety of programs and approaches. Any new tools or approaches such as self-suggestion, mental visualization, or meditation must be brought into your daily life and made to work for you. The initial excitement (which really comes from a fresh perspective on things) will die down and does not in itself guarantee any lasting changes. You can make the mistake of chasing after this rush of excitement and energy, when what is needed is to settle down and learn to apply what you have learned in your own life.

Any popular psychological program, and in fact any psychotherapy or counseling, must always be taken with a grain of salt and viewed as to whether it really works for you.

### Being Realistic

Taking a realistic approach is a matter of common sense. If you had a high school education and wanted a job as an LSI engineer, you would be very unrealistic. If you had a high school education but were working in a department doing LSI design and could show your superiors that you understood LSIs and their design, you might be able to get the engineering job. The second case is possible, ambitious, and not too unrealistic. Obviously you would have had to do a great deal of studying and learning.

If you had only a high school diploma and wanted to advance in the quickest fashion, it would be more realistic to pursue computer programming. The requirements are less rigid, and you can go far if you can demonstrate ability and aptitude for programming. However, interviewers are looking more and more for a college degree, particularly one in computer science. Also, what I have said is no secret, and many people are trying to move into computer programming. Fortunately, the long-term demand remains high, though in times of recession it may diminish.

### Be Positive

I have said much about this already. The important thing to remember is that most of us do not aim high enough. That does not mean that simply by being more ambitious or setting higher goals, success will come automatically. But it is true that what we accomplish follows what we think we can accomplish.

For instance, if you go for an interview convinced that you will not get a job, you will enter the interview in a depressed mood and probably seem unenergetic and uninteresting. Even if the interviewer liked your résumé he or she will observe how you act and conclude, "I am not sure I want this dull, discouraged person working for me."

If you enter the same interview feeling positive and that you have a good chance of getting a job you will appear brighter, more relaxed, and more interesting. You will probably get the job.

This goes back to our basic perceptions of the world and of a given situation. In many ways our actions are closely tied to our perceptions and attitudes.

Here is another example. If you want a favor from someone, you can approach the person in different ways. If you think the person is going to be obstinate and unhelpful, you may assume a demanding manner, already resentful about the expected rebuff. That approach will alienate the person, who probably will not help you or will do so in a grudging manner. You will say to yourself that your expectation was justified, that the person does not really want to help you.

If, however, you think the person will be helpful, your approach will probably be more relaxed and be simply a request for help rather than a demand. The person will be much more likely to help you.

Most people really do not learn to have positive expectations. Their expectations are limited. Frequently, what school systems and employers expect and need from people is not creativity or individuality, but a conforming attitude. We are encouraged to perform and be intelligent, but only in certain subject areas and in certain limited ways.

That is true to a certain extent of the high-tech companies. They are actually in a contradictory position: They need creativity and initiative to create new products and make advances in technical fields, yet that creativity must be channeled in directions that will eventually lead to marketable products.

The creativity of the human mind is a wonderful and awesome thing. Compared to animals and even to the most sophisticated computers, the human mind is impressive. Yet to survive and succeed in society, we must conform and in effect channel that creativity in certain directions. Ultimately we limit our own conception of our intelligence and creative ability.

Yet each of us has great reservoirs of intelligence and ability that we do not adequately tap. That is because of the great powers of the human mind. We see it even in the so-called idiot savants. At first glance they seem to be mentally deficient, yet many of them show amazing mental ability in some specific area—such as performing complex mathematical feats or showing great artistic or musical ability or photographic recall.

If we could tap more of our untapped abilities, each of us could handle jobs that we would have thought far beyond our capabilities.

Often in the high-tech industries that does happen. Because a company has a desperate need to fill a certain job, it hires someone without the proper credentials or promotes from within. According to a rigid view of credentials, this person should not be able to handle the job, yet through study and application he or she often succeeds at the job.

Another factor that affects our success is that we are all a little neurotic. We all have some area of insecurity that shows up in our daily lives as a lack of self-confidence (or may be masked by unrealistic over-confidence or arrogance). It may appear as some quirk or pattern of behavior that ties up our mental energy and does not help us. Any counseling or psychology program that lets us learn more about ourselves helps free up our mental energies and puts us more in touch with those untapped resources. Programs that enable us to be positive, to visualize ourselves positively, and to be more self-confident also help us, since a more positive attitude affects our actions, and our actions lead to better results.

# 20

# Recent Changes in High-Tech

Changes have been sweeping through the high-tech industry in recent years. Many of the changes revolve around new developments—everything from new types of computers to new kinds of software packages. Some of these products are used by office and factory workers around the world, helping them to work faster and with greater accuracy. Other high-tech developments have been used in Hollywood and seen by moviegoers everywhere. Films like "Terminator 2" and "Beauty and the Beast" have dazzled audiences with their computer-based special effects and animation. Changes like these have had a positive impact on the industry.

Not all of the industry changes, however, have been so positive. Although the 1990s have brought exciting new products and services, they also have brought instability and competition for computer companies. The industry has experienced competition both from within the United States and from foreign companies. On the hardware side of the business, for example, a number of smaller firms have taken on the big guys. Many computer companies that are not household names have taken a larger chunk of the market share, and the giant corporations have been forced to cut back or reorganize.

"People are more willing to buy from smaller companies," says one executive. "The continued recession put pressure on corporate computer buyers to justify their expenses. That forced, for the first time, large-scale buyers to consider less expensive, high-quality alternatives to the name brands."

Industry analysts say this instability is a sign that the high-tech field has matured. Although overall growth continues, it has slowed; and managers who do hiring are placing greater and greater emphasis on education and proper training for new employees. During the 1980s, a computerphile (someone totally engrossed in computer programming and computer technology) probably would have been able to land a high-paying job even without a college education. Today, however, the requirements for that same job are much stiffer. Many corporate managers are looking for college or technical school degrees or extensive training from their job applicants.

Substantial opportunities do exist to advance steadily in a high-tech career. To do this, you should take advantage of every training opportunity, whether it is taking a course at your current company, shifting to another company, or enrolling in classes at your local college or technical school.

## COMPETITION IN THE COMPUTER INDUSTRY

In the early 1980s, the computer industry faced recession, as did the rest of the country. The computer companies, however, seemed to hold up well and offer a haven of opportunity. Some companies certainly experienced layoffs and cutbacks, but hiring did occur and salaries continued to rise.

Toward the mid-1980s the computer industry entered a boom period of amazing proportions. The industry seemed to have endless promise, and hiring occurred at a frenzied rate. Companies typically grew at a rate of 15 to 25 percent a year or more.

Those were the years of the best opportunities in the high-tech field. You could be hired without having the best credentials and advance even if you did not have quite the "right" educational degree. The tremendous growth meant that good workers were desperately needed and competition among companies was intense. Thus, advancement was rapid and salaries rose quickly.

What has happened now, however, is that the computer industry has grown up and confronted reality. That reality is that no industry grows at a breakneck pace forever. The reason for much of the growth during the mid-1980s was that companies were paying enormous sums to computerize their offices and factories. Every one of them, it seemed, needed computers. Now all those companies have computers, and about 90 percent of consumer and corporate sales in the worldwide computer industry are for replacement computers. That accounts for the slowdown in hardware sales. Most consumers and corporations do not replace their computers every year.

In the previous period, it seemed that there was enough business for every company and all could grow rapidly together. Now there is much more competition. A company without the right mix of computer products or without advanced technology can suddenly find itself falling behind and its sales declining sharply. The hallmark of the current period could be called mixed growth: While some companies are doing well, others are floundering. Recently it has been the smaller companies that are doing well, largely because they offer quality products at lower prices. Companies like Zeos, Dell, and Microsoft saw more than 50 percent growth in revenue in 1991. The larger companies, such as

**151**

Amdahl, Wang, and Tandy, saw declines in revenue of more than 10 percent.

Many computer companies seem to have grown complacent, to have been caught with old technology in their products, or to have been insensitive to customers' desires and needs. On the other hand, smaller companies with innovative products have started up and grown rapidly. The message seems to be that a company that offers up-to-date technology and meets customer needs will do well.

Not only have computer makers been struggling with periods of poor revenues, but makers of computer peripherals also have seen slower growth. Although sales of laser printers are booming, declining prices are cutting into revenues. Another reason for lower earnings is growing competitiveness.

Many of the silicon chip manufacturing companies, mostly centered in California, have suffered a longer slump, partly because of competition from Japan. The competition in that area may be coming to an end, however, as American and Japanese chipmakers begin to form alliances. The world's biggest computer company, IBM, recently joined Toshiba and Siemens to develop a 21st-century chip. Toshiba is Japan's second-largest chipmaker, and Siemens is Europe's number three semiconductor house.

The semiconductor industry has seen other such collaborations. Texas Instruments is teamed with Hitachi, and Motorola Inc. has joined Toshiba, to develop memory chip technology. One of the reasons for these alliances is the astronomical cost of pursuing silicon chip development. The factories alone cost hundreds of millions of dollars.

Much has been made of the threat of competition from Japan in the computer industry, and Japanese companies have indeed had successes in the American market, particularly in peripherals such as printers and in other areas such as personal computers.

There is much discussion of the Japanese competitive spirit, technical know-how, and strong work ethic. The Japanese have shown themselves to be keen competitors in popular electronic products such as televisions and automobiles.

The Japanese, of course, deserve great credit. In certain areas their presence is and will remain strong. But overall the American computer industry has done very well against Japan and will continue to do so.

The reason is that the industry is highly competitive within the United States. The American industry is young, flexible, and vigorous compared to the popular electronics industry, which over the years became complacent and less concerned with quality. When the Japanese (and other Far Eastern competitors) arrived, the electronics industry was slow to respond.

The computer industry, however, has become very conscious of such matters as quality, for which the Japanese are known. In fact, the Japanese industry developed its modern concern with quality based on the theories of American quality-control leaders such as W. Edwards Deming. The Japanese were more energetic in applying these theories for a time, until American industry began to wake up.

What does this competition mean for you if you work in high-tech? Generally, it means that opportunities will abound, since competition for talented workers will be high. Companies will need hard-working, technically oriented employees to maintain a competitive position and will pay well for good people.

The computer industry, however, will be somewhat less stable. If your company starts slipping, you may be laid off, raises may be postponed, or you may be forced to take a pay cut.

Remember, however, that salaries in the computer industry tend to be high. If you receive a salary cut it can be distressing, but your pay may still be above those in other sectors of the economy. On the average, people in the computer industry continue to receive better raises than people in most other sectors. Being laid off from a job can also be upsetting, yet you have a good chance of being hired by another company in the computer industry.

Some of what has been said here may sound discouraging, but remember that at one time or another any industry can be affected by a slump. It also seems that a hallmark of work in modern America is a measure of insecurity and job change. It is difficult to find a sector of the economy that is a haven of security for the worker. In the 1950s and 1960s, some areas were held up as models of stability: civil service, tenured professorships at colleges, and positions at large corporations such as the old AT&T. Yet each of these areas has endured its share of problems over the years.

During the course of a career in high-tech, you may have to show some flexibility. You may need to work as a temporary contract worker for periods of time. You may need to take a job that is less than ideal, either to get into a desirable company or to stay with a company during a slump.

On the other side of the coin, the high-tech industry has generously rewarded those who have worked in it, with high salaries and wages, stock plans and options, and interesting and high-status jobs.

Every industry has its periods of ups and downs. Although during the past few years the computer industry has been in a slump, it is working to turn that around. Many high-tech companies now are concentrating on the small-business market and developing software. Both of these areas, especially software, offer hope for the high-tech industry to improve revenues and increase growth.

## ADVANCES IN THE COMPUTING INDUSTRY

One of the most exciting advances in the industry has been the joining of computer technology with other technologies to form multimedia. Multimedia systems are the merging of personal computers, televisions, and compact disk players. An example is a system that plays conventional five-inch music CDs and can be hooked up to almost any stereo. You also can plug the system into your television set and play CD-like disks that contain interactive video.

With a system like this, instead of just watching a video about the Smithsonian Institution, you can visit the 150 exhibits in whatever order you like. You interact with the video, telling it what you want to see.

Another development is called MPC (multimedia personal computer). The MPC is a functional computer on which you can run interactive programs. Some of the CD-ROM programs combine articles, images, and sound. So now you can learn about whales by using your MPC. Not only can you read several articles, but you can also view photos, hear the whales communicate, and even watch an animated film of whales swimming.

Other advances are being made in the category of gadgets. One such gadget is called a Personal Digital Assistant, or PDA. PDAs are hand-held personal computers that include an electronic datebook, a Rolodex, a notepad, and a fax machine. You can do calculations, list phone numbers, and maintain schedules and lists of things to do. The PDA can even send a fax. The user simply writes on the computer screen, "Fax to Jefferson," and the PDA automatically looks up his fax number and sends the message.

Other developments include a "personal communicator" that lets you send written notes and a hand-held computer that "reads" hand printing. Electronic book players are expected to be on the market soon. These hand-held machines read CD-ROM disks that store novels, textbooks, and even entire encyclopedia sets.

In the area of computerized design, work is being done to develop 3-D printing. Three-dimensional printing is made possible by the use of computers and robots. A computer image guides a laser beam over a vat of light-sensitive liquid plastic. Using this, engineers can create a solid replica of their designs within hours, cutting weeks off the production time of automobiles or airplanes. In the future, companies may even be able to "fax" a solid object to distant locations.

An area that has been revolutionized by the power of the computer is desktop publishing. Desktop publishing involves using a personal computer for the writing, layout, and printing of documents that normally would have required a typewriter and professional printing facilities. With desktop publishing, any company or individual can lay out a finished newsletter or book and print large numbers of professional looking copies at a moderate cost.

With these new products comes the need for well-educated, highly qualified high-tech workers. Companies will be looking for software engineers to continue to create new products. They will be looking for draftsmen and electrical, mechanical, and manufacturing engineers to design and build the hardware. High-tech firms also will need customer service representatives, technical writers, and user support persons to assist buyers in understanding and using their purchases.

Another major advance in computers is artificial intelligence (AI) computing and programming. AI sounds a little like science fiction and may conjure up images of "intelligent" computers replacing humans in jobs and taking over the world. For the near future, however, at least until the year 2,000, AI should be regarded merely as a different kind of computing approach used for particular problems. AI is a branch of computer science that deals with using computers to simulate human thinking such as reasoning and communicating. It also tries to mimic biological senses, including sight and hearing. But the AI industry is a long way from producing any computer that will duplicate the capabilities of the human brain.

AI often uses special computer languages such as LISP and PROLOG and sometimes uses special hardware.

One of the most common AI approaches is the "expert system," which is a software program containing a large number of rules and data in a particular problem area, such as medical diagnosis. (Such systems are not meant to replace doctors but to assist them with decision-making.) Expert systems also are useful to managers in the areas of finance and manufacturing, helping them to identify future trends.

Other AI advances include psychological analysis programs that measure human traits like extroversion, kindness, nervousness, and intelligence. These may be used to select an ideal juror for a case or to evaluate prospective salespeople.

Optical recognition and speech recognition are two more areas of AI. In optical recognition, a computer "reads" a printed document and turns it into computer text. Speech recognition includes giving the computer spoken commands. For instance, if you are working in a manufacturing operation you may give a computer commands by speaking to it. The computer in turn controls some piece of machinery. Speech recognition is used quite often by the blind or those unable to use

**157**

their hands, enabling them to write letters or computer programs.

In general, AI can make computers easier to use by operating in ways that are more natural for the user. Presently you are required to use very specific commands. An AI "interpreter," however, can accept a wide variety of commands. That is more like the way you usually ask someone to do something; you use different words and phrases to convey the same idea.

AI has been overpublicized in recent years, and too much has been expected of it too fast. It may seem intimidating, sophisticated, and exotic. Keep in mind, though, that AI is only another form of computing whose techniques and approaches can be learned just like any other.

The greatest opportunities in AI will be for computer engineers. This area requires great technical ability and a lot of hard work to earn an advanced degree such as a master's or doctorate. You may spend a lot of time applying AI techniques to relatively ordinary problems, rather than working on something sophisticated and exciting.

Another advance in computing is the proliferation of individual computers, both low-cost personal computers and somewhat more expensive desktop workstations. This trend, which began in the 1970s, has grown explosively throughout the 1980s and early 1990s as advances in hardware make small computers cheaper and more powerful. It will undoubtedly continue with even greater strength during the mid- to late-1990s.

Today, you are more likely to have an individual computer on your desk, whereas in the past you would have had a computer terminal connected to a mainframe. The individual computer probably will be connected to a network of personal computers and larger computers.

Various types of computer administrators and consultants may be available to help you run your individual computer, but that kind of assistance varies greatly from one job situation to another. You may have to take more responsibility for your individual computer than if you just had a terminal. This can include calling for service when needed and doing things like loading a software update from a disk. As in many other situations, if you learn what needs to be done for the computer, you will be regarded as an asset to your group.

Having your own computer can also be a great asset in that you are not dependent on a mainframe that may be overloaded with other work. You may be able to experiment more and learn more about computers. Some companies even may allow you to do a project with the computer that is not directly work-related.

Another trend that is continuing in the 1990s is the proliferation of users' groups and computer magazines. Users' groups usually are oriented around a particular model of computer; they hold regular meetings and are sources of advice on problems you may have with your computer. These organizations are particularly valuable if you work at home, run your own small business, or work in a very small company.

### Prerequisites, Training, and Advancement

The computer industry has become much more credential-oriented in the 1990s. You need the proper credentials to get a good job. In almost all cases, this means training or education beyond high school, and the more the better.

More colleges are offering training programs, certificates, or associate degrees in areas of computer technology. The certificate or degree that is appropriate depends on your area of interest.

Technical writing is an area that has become more credentials-conscious since I began in the field a dozen or so years ago. My only qualification was a bachelor's degree. I probably was in the last "generation" not to have specific training in technical writing.

Many kinds of training programs have developed. The most common one seems to be a nine-month program (for people who already have a bachelor's degree but not necessarily a technical one) at an accredited college or university. If you do well in such a program and find a good internship, you should be able to get a good starting job as a technical writer.

If you cannot immediately obtain a certificate or a degree, consider part-time or night-school programs.

Many computer companies have training programs for employees. Some companies may give you a leave of absence to attend college and even pay your costs to do so. The requirement here is that you get the job to begin with and then prove yourself capable.

A degree in computer science can be valuable no matter what your area of interest. This is true in marketing, training, customer service, computer repair, programming, and technical writing.

# Appendix
# Computer Associations and Societies

The following list of computer organizations and societies may prove useful to students and those just entering the computer field. You can write or phone for membership information and a list of publications.

American Society for Information Science
8720 Georgia Avenue
Silver Spring, MD 20910-3602
(301) 495-0900

Association of Computer Professionals
230 Park Avenue
New York, NY 10169
(212) 599-3019

Association of Computer Users
PO Box 2189
Berkeley, CA 94702-0189
(415) 549-4336

Association for Computing Machinery
11 West 42nd Street
New York, NY 10036
(212) 869-7440

Association for Systems Management
1433 West Bagley Road
PO Box 38370
Cleveland, OH 44138
(216) 243-6900

Association for Women in Computing
PO Box 21100
St. Paul, MN 55123
(612) 681-9371

Computer-Aided Manufacturing International
1250 East Copeland Road
Arlington, TX 76011
(817) 860-1654

Computer Dealers and Lessors Association
1212 Potomac Street NW
Washington, DC 20007
(202) 333-0102

Computer and Communications Industry Association
666 11th Street NW
Washington, DC 20001
(202) 783-0070

Computer and Automated Systems Association of the
Society of Manufacturing Engineers
Box 930
1 SME Drive
Dearborn, MI 48121
(313) 271-1500

Data Entry Management Association
101 Merritt 7
Norwalk, CT 06851
(203) 846-3777

Data Processing Management Association
505 Busse Highway
Park Ridge, IL 60068
(708) 825-8124

"intelligent" computers replacing humans in jobs and taking over the world. For the near future, however, at least until the year 2,000, AI should be regarded merely as a different kind of computing approach used for particular problems. AI is a branch of computer science that deals with using computers to simulate human thinking such as reasoning and communicating. It also tries to mimic biological senses, including sight and hearing. But the AI industry is a long way from producing any computer that will duplicate the capabilities of the human brain.

AI often uses special computer languages such as LISP and PROLOG and sometimes uses special hardware.

One of the most common AI approaches is the "expert system," which is a software program containing a large number of rules and data in a particular problem area, such as medical diagnosis. (Such systems are not meant to replace doctors but to assist them with decision-making.) Expert systems also are useful to managers in the areas of finance and manufacturing, helping them to identify future trends.

Other AI advances include psychological analysis programs that measure human traits like extroversion, kindness, nervousness, and intelligence. These may be used to select an ideal juror for a case or to evaluate prospective salespeople.

Optical recognition and speech recognition are two more areas of AI. In optical recognition, a computer "reads" a printed document and turns it into computer text. Speech recognition includes giving the computer spoken commands. For instance, if you are working in a manufacturing operation you may give a computer commands by speaking to it. The computer in turn controls some piece of machinery. Speech recognition is used quite often by the blind or those unable to use their hands, enabling them to write letters or computer programs.

In general, AI can make computers easier to use by operating in ways that are more natural for the user. Presently you are required to use very specific commands. An AI "interpreter," however, can accept a wide variety of commands. That is more like the way you usually ask someone to do something; you use different words and phrases to convey the same idea.

AI has been overpublicized in recent years, and too much has been expected of it too fast. It may seem intimidating, sophisticated, and exotic. Keep in mind, though, that AI is only another form of computing whose techniques and approaches can be learned just like any other.

The greatest opportunities in AI will be for computer engineers. This area requires great technical ability and a lot of hard work to earn an advanced degree such as a master's or doctorate. You may spend a lot of time applying AI techniques to relatively ordinary problems, rather than working on something sophisticated and exciting.

Another advance in computing is the proliferation of individual computers, both low-cost personal computers and somewhat more expensive desktop workstations. This trend, which began in the 1970s, has grown explosively throughout the 1980s and early 1990s as advances in hardware make small computers cheaper and more powerful. It will undoubtedly continue with even greater strength during the mid- to late-1990s.

Today, you are more likely to have an individual computer on your desk, whereas in the past you would have had a computer terminal connected to a mainframe. The individual computer probably will be connected to a network of personal computers and larger computers. Various types of computer administrators and con-

Desktop Publishing Applications Association
c/o Herbert Communication
3 Post Office Road
Waldorf, ND 20602-2710

Hispanic Computing Association
2388 Mission Street
San Francisco, CA 94110
(415) 824-8337

Independent Computer Consultant Association
933 Gardenview Office Parkway
St. Louis, MO 63141
(314) 997-4633

Institute for Personal Computing
PO Box 558250
Miami, FL 33255
(305) 577-8394

National Association of Computer Consultant Businesses
1250 Connecticut Avenue NW
Washington, DC 20036
(202) 637-6483

National Association of Desktop Publishers
1260 Boylston Street
Boston, MA 02205
(617) 426-2885

National Association of Professional Word Processing
Technicians
110 West Bayberry Road
Philadelphia, PA 19116
(215) 698-8525

National Computer Graphics Association
2722 Merrilee Drive
Fairfax, VA 22031
(703) 698-9600

National Society for Computer Applications in
Engineering, Planning, and Architecture
c/o Robert D. Marshall
Edwards and Kelcey Inc.
705 South Orange Avenue
Livingston, NJ 07039
(201) 994-4520

National Systems Programmers Association
4811 South 76th Street
Milwaukee, WI 53220
(414) 423-2420

Special Libraries Association
1700 18th Street NW
Washington, DC 20009
(202) 234-4700

Women in Information Processing
Lock Box 29173
Washington, DC 20016
(202) 328-6161

World Computer Graphics Association
2033 M Street NW
Washington, DC 20036
(202) 775-9556

# Glossary

**access code** Unique combination of letters or numbers used to gain access to a computer network or online service. Sometimes called **user name** or **user ID and password**.

**active file** File in use that any currently issued command will affect.

**add-on** Device, such as an external hard disk, that will expand the capacity and capabilities of a computer system.

**algorithm** Set of rules or instructions that can be followed to carry out a particular task. A recipe in a cookbook, for example, could be considered an algorithm.

**algorithmic language** Any programming language that focuses on problem-solving by using algorithms. BASIC, C, FORTRAN, and PASCAL are examples of algorithmic languages.

**alphanumeric** Descriptive of a data field that can hold both letters and numerals.

**analog computers** Computers typically used for scientific and industrial operations that do not store data as digital bits but instead represent variable quantities (such as temperature) with a proportionally variable current or voltage.

**application** Computer program designed to help perform a certain type of work, such as payroll, order entry, or word processing. An application can manipulate text, numbers, graphics, or a combination of those elements.

**application program** Single, user-written program designed to accomplish a specific user task.

**architecture**    General term referring to the structure of all or part of a computer system.

**arithmetic logic unit (ALU)**    Hardware in the computer containing the circuits and memory that perform arithmetic, comparative, and logical functions.

**artificial intelligence (AI)**    Branch of computer science that deals with programming computers to process information like humans, using deduction, inference, and the ability to learn from experience.

**ASCII (pronounced "askee")**    Acronym for American Standard Code for Information Interchange, a standardized code that enables computers and computer programs to exchange information.

**assembly language**    Low-level, machine-oriented programming language in which each statement corresponds directly to a single machine instruction. It is difficult and time-consuming to code but produces very fast and efficient programs.

**audit trail**    The series of printed computer reports that are produced during processing to show that no errors or omissions have occurred from first input to final output.

**backup**    The process of storing a duplicate copy of files or programs on a disk or magnetic tape, to be available if the original is damaged or erased.

**BASIC (*B*eginner's *A*ll-purpose *S*ymbolic *I*nstruction *C*ode)**    The most popular high-level, English-like programming language; it is often taught to beginning programmers because it is easy to use and understand.

**batch**    Group of documents or data records that are processed as a unit.

**batch processing**    On microcomputers, the running of a batch file, a stored "batch" of operating system commands carried out one after the other without intervention. On larger computers, batch processing involves acquiring programs and data from users, running them one or a few at a time, and providing the results.

**baud** Unit of measurement for the transmission speed of data over telephone lines.

**binary number system** The standard number system used on all digital computers. It uses only the digits 0 and 1, which can be represented as the OFF and ON states of a bit of electronic memory.

**bit** Short for binary digit, the smallest unit of information processed and stored by a computer. It is either a 0 or a 1 in the binary number system.

**block** Multiple data records handled as a single group for efficient transfer into and out of memory.

**branch** Programming technique in which control is passed to an instruction that is not the next sequential instruction in the program.

**buffer** Temporary storage area for data that helps to compensate for the difference in speed between mechanical and elctronic devices in the computer's hardware configuration.

**bug** Error in software or hardware. In software, a bug causes a program to malfunction or to produce incorrect results. In hardware, a bug is a recurring physical problem that prevents a system or set of components from working together properly.

**byte** Unit of information containing 8 bits. It is the amount of memory used to store a single character, such as a letter, a numeral, or a punctuation mark.

**C** Structured programming language that is considered the closest thing to a standard programming language in the microcomputer/workstation marketplace.

**CAD** (*computer-aided design*) Programs used in designing engineering, architectural, and scientific models, ranging from tools to buildings to airplanes.

**calculator** Device that performs arithmetic operations on numbers but requires continual interactive instructions from the operator.

**CAM** (*computer-a*ided *m*anufacturing) Manufacturing machinery that is controlled by a computer.

**cardreader** Input device. Magnetic card readers read information that has been magnetically encoded on a plastic card, like a credit card. Punched-card readers read computer data from punched cards.

**cathode-ray tube (CRT)** Large electronic vacuum tube, similar to a television tube, used to display information or data quickly on a microcomputer screen.

**CD-ROM (compact disk read-only memory)** Form of high-capacity data storage that uses laser optics for reading data.

**central processing unit (CPU)** The electronic "brain" of the computer, which interprets and executes instructions.

**chip** A small part that contains electronic circuitry.

**clone** A copy. Clones are look-alike, act-alike computers that contain the same microprocessors and run the same programs as better-known, more prestigious, and more expensive machines.

**COBOL** (*C*ommon *B*usiness *O*riented *L*anguage) High-level, English-like computer language that is widely accepted and used, expecially in business applications.

**code** Term for computer instructions.

**communication** Transmission of data or information between remote locations.

**compile** To translate all the source code of a program from a high-level, English-like language into a lower-level, machine-readable format. A program that performs this task is known as a compiler.

**computer** Any machine that accepts structured input

data, performs calculations and logical operations on that data, and reports the results.

**computer-output microfilm (COM)** Microfilm that can record data from a computer without the use of a camera for efficient storage of high-volume reports.

**computer program** Set of instructions in a computer language to be executed on a computer to perform a specific task.

**computer science** The study of computers, including design, operation, and use.

**console** The terminal used by the computer user for interactive communication with the computer.

**control section** The part of the central processing unit that decodes each programmed instruction in proper sequence.

**control total** The printed sum of a particular field in a group of records, accumulated to ensure the accuracy of the computer processing. It is part of the audit trail.

**control unit** Device or circuit that regulates between the computer and one of the peripheral input/output devices.

**core-image library** A portion of the master system disk library that contains programs in machine-language executable form.

**crash** The sudden failure of either a program or a disk drive.

**data** Any collection of characters—numeric or alphabetic—that can be assigned a meaning and processed by a computer to produce meaningful information.

**database** File consisting of related, integrated records together with a collection of operations that facilitate searching, sorting, recombination, and other activities.

**database management system (DBMS)** Set of software programs for creating, modifying, updating, and retrieving information from a database system.

**data entry**   The process of writing new data into computer memory, typically from a keyboard.

**data file**   File consisting of data—text, numbers, or graphics—as distinct from a program file of executable instructions.

**data processing (DP)**   Work performed by computers: the inputting of data, the application of a series of changes leading to a result, and the reporting of that result in the form of useful information.

**data transfer**   The movement of information from one location to another, either within a computer or between a computer and an external device (as between two computers).

**debug**   To remove the errors (bugs) from a computer program.

**decimal numbering system**   Numbering system using the base 10 and the digits 0 through 9.

**desk checking**   Tracing the various paths though a program to verify the logic before submitting it for a computer test.

**desktop publishing**   Use of a computer and specialized software to combine text and graphics to create a document that can be printed on a laser printer or a typesetting machine.

**digital computer**   The dominant variety of computer for both scientific and business applications. Data are stored in a digital computer in binary form, as a series of 1s and 0s.

**direct access**   The ability of a computer to go straight to a particular storage location in memory or on a disk and retrieve or store an item of information. Also called **random access**.

**disk**   Device for storing information used by a computer.

**disk pack** High-volume storage device consisting of a collection of disks in plastic housing. Data are stored magnetically on the surface of the disk, which is used both for input and output.

**display** The visual output device of a computer.

**distributed processing** Form of information processing in which work is performed by individual computers that are linked through a communication network.

**documentation** The flowcharts, logic narratives, truth tables, and other written instructions that explain the functioning of a computer system or program.

**dump** An automatic printout of main memory that occurs when a program fails during processing. The dump will assist the programmer who must debug the program logic.

**electronic data processing (EDP)** *See* **data processing**.

**electronic funds transfer (EFT)** The transfer of "magnetic money" over phone lines from one bank to another.

**electronic mail** The transmission of messages over a communication network. Also called **e-mail**.

**encode** To change alphabetic and numeric characters into computer code by setting various patterns of bits to 0 or 1. Some common codes are octal, hexadecimal, and binary.

**field** Location in a record in which one data item, made up of one or more characters, is stored.

**file** Collection of related records. For example, if a name, address, and phone number make up a record, the phone book would be a file.

**file maintenance** The process of using a program to alter a file by inserting, deleting, changing, or copying.

**first-generation computers** The first electronic computers, characterized by the use of vacuum tubes.

**flowchart**   Chart that shows the movement of data through a program or within a computer system. A flowchart uses agreed-upon symbols such as squares, diamonds, and ovals to describe various operations pictorially.

**FORTRAN** (*Formula Translation*)   The first high-level, English-like programming language. It is used heavily in scientific and engineering fields.

**freeware**   Computer program given away free of charge, often through computer bulletin boards or user groups.

**general-purpose language**   Programming language like BASIC or C, designed for a variety of applications.

**hacker**   Person engrossed in computer programming and computer technology.

**hardware**   The physical computer and related equipment, including printers, modems, and mice.

**hexadecimal**   Numbering system with the base of 16 that consists of digits 0 through 9 and the upper- or lower-case letters A through F.

**high-level programming language**   English-like language that makes programming easier but must be translated by a compiler program into machine language the computer understands. Examples include COBOL, FORTRAN, and BASIC.

**home computer**   Personal computer designed for home use, usually available for a lower price than an office PC.

**housekeeping**   The functions performed in the first paragraph of a high-level program, such as setting counters to zero, opening files, and checking labels.

**hypertext**   Medium in which images, sounds, and actions are linked together in such a way that a user can browse through related topics in any order.

**information**   Processed data in a usable form.

**input**   Data entered into a computer for processing.

**input/Output (I/O)**   The tasks of gathering data for the computer to process (usually entering with a keyboard or mouse) and making the results available to the viewer (usually with a display or printer).

**instruction set**   Group of machine instructions in a given computer language that a microprocessor recognizes and can execute.

**integrated circuit (IC)**   Highly miniaturized electronic chip that contains thousands of transistors and other electronic components.

**intelligent terminal**   Terminal having its own memory and microprocessor.

**interactive program**   Program that interacts with the user who uses a keyboard, mouse, or joystick to provide responses to the program. A computer game, for example, is an interactive program.

**job**   In computer programming, one program that the computer must run.

**job control language (JCL)**   Special programming language that controls the computer's operating system as it executes application programs.

**job step**   Single-application program, usually one of many executed sequentially in the jobstream.

**joystick**   Pointing device used to play computer games and for other tasks.

**K**   Abbreviation for kilo-. In computing, K stands for 1024 characters of main storage (often rounded to 1000). For example, 64K would be read as 64,000 bytes of memory.

**key**   The field in a record used to identify the record, to retrieve it directly from a disk, and, sometimes, to sort it. In payroll application, employees social security numbers often are used as keys.

**keyboard**   Part of a computer that resembles a typewriter and is used to input data.

**keypunch** Device used to punch data holes into paper cards to provide data for early computers.

**label** Word, symbol, or group of characters used to identify a file, a storage medium, or an element defined in a computer program.

**language processor** Computer program that accepts instructions written in a particular computer language and translates them into machine language. Also called a **compiler, interpreter,** or **assembler.**

**leased line** Dedicated or leased telephone channel for private use. A leased line is faster, quieter, and generally more expensive than a standard switched line. Leased lines usually are used for data communications.

**light pen** Light-sensitive pointing device that can be attached to a computer. The user can hold the wand up to the screen to select items or choose commands instead of typing selections on a keyboard.

**line printer** Large high-speed printer used for printing out computer information.

**linkage editor** (or **linker**) Manufacturer-supplied program that links data files and compiled modules to create an executable program.

**low-level language** Language that has a close correspondence to the computer's instruction set, such as machine language or assembly language.

**machine code** (or **machine language**) The only language computers understand. High-level languages like BASIC or PASCAL are ways of structuring the human language so people can program computers to perform specific tasks. A compiler translates these high-level languages into machine code.

**magnetic core** (or **core**) Early type of computer memory. A core consisted of small iron circles strung on control wires and magnetized to represent either binary 0 or 1. Now the term is sometimes used to refer to main memory of any computer system.

**magnetic disk (or computer disk)** Flat, circular metallic plate coated with a material that can be magnetized. The two main types are hard disks and floppies.

**magnetic-ink character recognition (MICR)** A reliable method of character recognition that reads characters printed with magnetically charged ink. An example of magnetically charged characters are the numbers at the bottom of bank checks.

**magnetic tape (or tape)** A reel of tape with an oxide surface that can be magnetized to store data. On personal computer systems, ordinary two-reel audio cassettes often are used for this purpose.

**mainframe computer** High-level computer designed for complex and intensive computational tasks. The most powerful mainframes, called supercomputers, are used heavily by researchers in science, big business, and the military.

**main storage (or primary storage)** The central general-purpose storage region of a modern digital computer. Secondary storage options include both disks and tapes.

**management information system (MIS)** Group of business software systems designed to assist management in decision-making and long-range planning.

**mass storage** Generic term for disk, tape, or optical disk storage of computer data. Large amounts of data can be stored on these secondary storage devices in comparison with computer memory capacity.

**master file** Large file stored on disk or tape that is updated periodically (usually weekly or monthly) and reflects the ongoing financial history of a company.

**megabyte (MB)** About one million bytes of memory or storage (exactly 1,048,576 bytes).

**microcomputer** Computer built around a single-chip

microprocessor; less powerful than minicomputers or mainframe computers.

**microprocessor** A central part of the computer; the central processing unit that has been miniaturized so that it will fit on one chip.

**minicomputer** Mid-level computer built to perform complex computations, usually used in business or scientific fields.

**modem** (*mo*dulator/*dem*odulator) Device used to transmit electronic data over telephone lines.

**modular programming** Technique that involves dividing a large computer program into simple functional units. This speeds up the programming process by allowing several programmers to work on a a single program at the same time.

**monitor** The screen on which images generated by the computer's video adapter are displayed.

**mouse** Common pointing device used to select items or choose commands on the computer screen. The user presses one of the mouse's buttons, producing a "mouse click" to make a selection.

**multiplexing** Electronic process that allows two or more data transmissions to share a telephone line.

**multiprocessing** The linking of two computer systems to allow the processing of two programs simultaneously. Each processing unit works on a different set of instructions.

**multiprogramming** Execution of two programs at the same time on one central processing unit by rapidly switching back and forth between them.

**nanosecond (NS)** One billionth of a second; a time measure used to represent computing speed.

**natural language** Human languages, as opposed to programming or machine languages.

**network** Group of computers and other devices connected by cables, telephones, or other communication links.

**number crunching** Programs or computer applications that involve numerous mathematical calculations and often require large compilers and large amounts or computer time.

**numbering system** System for the writing and manipulation of numbers using agreed-upon symbols and rules. Examples include decimal, binary, octal, and hexadecimal numbering systems.

**object code** The complete machine-language program that can be directly executed by a computer's central processing unit.

**octal numbering system** Numbering system using a base of 8 and the digits 0 through 7.

**offline** State in which a device, such as a printer, cannot communicate with or be controlled by a computer.

**online** State in which a device is activated and ready for operation, capable of communicating with or being controlled by a computer.

**online direct access system** Data processing operation controlled by the main computer with files stored on disk, using the direct-access method for high-speed retrieval.

**operating system (OS)** The manufacturer-supplied master program responsible for controlling the allocation and usage of hardware resources, such as memory, central processing unit time, disk space, and peripheral devices. Popular operating systems include MS-DOS, the Macintosh OS, OS/2, and UNIX.

**operation code** (or **Opcode**) The part of an assembly language or machine language instruction that tells what operation is to be performed.

**optical character recognition (OCR)** Process of

examining printed characters—such as newspaper articles—and determining their shapes by detecting light and dark. Once the shapes are determined, they are translated into computer text.

**original equipment manufacturer (OEM)** Firm that incorporates the products of another company in its retail product. Also, a software house that adds specialized software to a minicomputer and retails a complete turnkey system.

**output** The results of computer processing.

**page** A fixed-size block of memory.

**paging** Technique of implementing virtual storage. The virtual address space is divided into fixed-size blocks of memory (pages). These pages can be mapped onto any of the physical addresses available on the system.

**parallel interface** Hard-wired interface between two devices that transmits all bits in a byte at the same time.

**parallel processing** Computer processing approach that solves a problem by using several processing units to work on it at the same time. These units can be combined in the same computer or can be in separate computers connected by a network.

**parity** Error-checking procedure that verifies the accuracy of data by adding the number of bits that are set ON (the binary 1s) in each byte.

**PASCAL** A concise, structured programming language that gained popularity during the 1980s as a standard development language on microcomputers.

**peripheral** Term for devices like disk drives, printers, modems, and joysticks that are connected to and controlled by a computer.

**personal computer (PC)** Computer designed for use by one person at a time.

**PL/I** (*Programming Language I*) Language developed by IBM to combine the best features of COBOL, ALGOL, and FORTRAN, while introducing new concepts. The result was a structured language so complex it met with only limited acceptance. It is still used, however, in some academic and research environments.

**plotter** Output device that can produce graphic drawings, including charts and diagrams, using either pens or electrostatic charges and toner. Pen plotters draw on paper with one or more colored pens. Electrostatic plotters "draw" a pattern of electrostatically charged dots on papers and then apply toner to fuse it into place.

**plug-compatible manufacturer** Producer of peripherals that can be easily plugged into a popular main computer.

**point of sale** (**POS**) The technique of using computerized transaction systems (or online cash registers) to capture sale and inventory data at the moment of sale.

**portable computer** Any computer designed to be moved easily, ranging from less than 2 lbs to about 30 lbs. Types include transportable, laptop, ultralight, and hand-held.

**printed-circuit board** A key part of the computer, a flat board of rigid fiberglass. Boards vary greatly in size. Components such as chips, resistors, and capacitors are mounted on the board.

**printer** Output device that puts text or a computer-generated image on paper or on another medium, such as a transparency.

**private line** *See* **leased line**.

**program** Series of instructions that can be executed by a computer.

**program maintenance** Modification of an operational program for the purpose of correcting errors in logic or for updating constants in the program to meet changing government regulations or business practices. An example would be changing the amount of federal taxes deducted from employees' paychecks in a payroll program.

**programmer** Individual who writes and debugs computer programs.

**random access memory (RAM)** Semiconductor-based memory that can be read or written by the microprocessor or other hardware devices.

**read-only memory (ROM)** Semiconductor-based memory that contains instructions or data that can be read but not modified.

**real storage** The amount of random access memory (RAM) storage in a computer, as opposed to virtual storage. Also called **physical storage** or **physical memory**.

**real-time processing** The processing of data so quickly through online techniques that the computer output accurately reflects the current state of the data in the real world. An example of this is stock prices, which change continually throughout the trading day. Some computers are designed and programmed to give users up-to-the-second stock prices.

**record** Collection of related fields or data items.

**register** A small, high-speed part of memory in the computer used to hold specific pieces of data related to activities going on within the system.

**remote terminal** Terminal at a distant location from the main computer to which it is attached. Remote terminals are connected to main computers by telephone lines.

**report generator** Program that uses a report "form"

created by the user to lay out and print the contents of a database.

**response time** The time between input and output in an online system.

**robot** Machine that can sense and react to input and cause changes in its surroundings with some degree of intelligence. Robots rarely are human-like in appearance but often mimic human movements in carrying out their work.

**second-generation computer** Computer that uses discrete transistors as its basic circuitry element.

**secondary storage** Any data storage other than a computer's random access memory, usually a tape or disk.

**segmentation** The technique of dividing a large program into pieces in order to run it on a small computer.

**semiconductor** A substance—usually silicon or germanium—that ranks between a conductor and a nonconductor in its ability to conduct electricity.

**sequential processing** Processing in the order in which the items of information are stored or input.

**service programs** Manufacturer-supplied utility programs that assist in the preparation or execution of user-written programs. Examples are the compiler, the assembler, the linkage editor, and the sort/merge.

**software** Computer programs.

**solid logic technology** Technique of hardware construction introduced by IBM in the third generation of computers, characterized by the use of integrated circuits to achieve extremely high circuit density.

**sort/merge program** Manufacturer-supplied program to sort or merge files by whatever key, field, and sequence the user specifies.

**source code** Human-readable program statements.

**source program** The human-readable program written by the programmer that will be input to the compiler.

**speech recognition** The ability of a computer to understand spoken words for the purpose of receiving commands and data input from the speaker.

**spooling** Technique of storing data in a disk or memory buffer until the printer is ready to process it.

**storage** Device on which computer data can be kept.

**structured programming** Technique that produces programs with clear design according to a top-down, modular scheme. Structured programming makes program maintenance easier and makes the program less difficult to read.

**supercomputer** Large, extremely fast, and very expensive computer used for complex calculations.

**superconductor** Substance that has no resistance to flows of electricity.

**supervisor** Another name for **operatng system.**

**system flowchart** Large pictorial diagram that shows the relationship between programs in a computer system.

**systems analyst** Person who works with users to design and develop computer systems.

**tape (magnetic)** Tape on which computer data are stored; it is wound on spools and looks like recording tape.

**telecommunications** The transmission of information —data, television pictures, sound, facsimiles, etc.— over telephone lines or cables.

**teleprocessing** The use of computers and communications equipment to access computers and computer files located elsewhere.

**terminal** Device with a keyboard that allows the user to communicate with the computer.

**testing** The preliminary running of a program with

test data to ensure that it is functioning properly before putting that program into live production.

**third-generation computer** Computer based on integrated circuits rather than on separately wired transistors.

**thrashing** The wasteful swapping of pages in and out of virtual storage rather than executing applications.

**throughput** The data processing rate in a computer system.

**time-sharing** The use of a computer system by many individuals at the same time. Users work at various terminals, and the computer system runs their separate programs concurrently.

**top-down design** System or program planned in a functional manner so that the most important functions are handled first and the highest level of detail is handled last.

**turnaround time** Elapsed time between the submission of computer input and the production of output.

**turnkey system** Finished system with all the hardware, software, and documentation installed and ready to be used.

**utility program** Program designed to perform maintenance work on the system or on system components; for example, a file recovery program.

**virtual machine** The use of systems software to make one computer system perform like a different computer system.

**virtual storage** Technique of using disk space to supplement semiconductor memory.

**voice input** Vocal instructions translated by a computer into commands it can execute.

**word length** The number of characters in a single computer word, typically 8, 16, or 32. Word length determines the limit of addressable memory.

**word processing (WP)** Use of an application program for preparation of textual materials like letters, reports, newspapers, magazines, and books.

**workstation** Desktop computer, with good graphics abilities, more powerful than a PC. Workstations often are used for graphics applications such as printed-cirucit board design and CAD.

# For Further Reading

### Computer Dictionary
Sippl, Charles J. *Computer Dictionary*, 4th ed. Indianapolis: Howard W. Sans, 1985.
Spencer, Donald. *Computer Dictionary*, 3rd ed. Merrifield, VA: Camelot Publishers, 1991.

### Computer Theory
Deem, Bill R.; Muchow, Kenneth; and Zeppa, Anthony. *Digital Computer Circuits and Concepts*, 3rd ed. Reston, VA: Reston Publishing Co., 1980.
Hall, Douglas V. *Digital Circuits and Systems*. New York: McGraw-Hill Publishing Co., 1989.
Streib, William J. *Digital Circuits*. South Holland, IL: Goodheart-Wilcox Co., 1989.

### General
Kidder, Tracy. *The Soul of a New Machine*. New York: Avon Books, 1981.
Lebow, Irwin. *Digital Connection: A Layman's Guide to the Information Age*. W.H. Freeman & Co., 1990.
Trainor, Timothy N. *Computers!* 2nd ed. New York: McGraw-Hill Publishing Co., 1989.

### Career Evaluation
Bolles, Richard. *What Color Is Your Parachute? 1991*, rev. ed. Berkeley: Ten Speed Press, 1990.
Masterson, Richard. *Exploring Careers in Computer Graphics*, rev. ed. New York: Rosen Publishing, 1990.
*Occupational Outlook Handbook, 1991–1992*. Washington, DC: U.S. Government Printing Office, 1991.

Spencer, Jean. *Exploring Careers in the Electronic Office,* rev. ed. New York: Rosen Publishing, 1989.

Spencer, Jean. *Exploring Careers in Word Processing and Desktop Publishing.* New York: Rosen Publishing, 1990.

### Résumé Writing

Beatty, Richard H. *Résumé Kit,* 2nd ed. New York: John Wiley & Sons, Inc., 1991.

Bostwick, Burdette E. *Résumé Writing: A Comprehensive How-to-Do-It Guide,* 4th ed. New York: John Wiley, & Sons, Inc., 1990.

Brennan, Lawrence D.; Strand, Stanley; and Gruber, Edward C. *Résumé for Better Jobs,* 3rd ed. New York: Monarch Press, 1987.

Lewis, Adel. *How to Write Better Résumés.* Woodbury, NY: Barron's Educational Series, 1983.

Reed, Jean. *Résumés That Get Jobs,* 5th ed. New York: Arco Publishing Inc., 1990.

### Computer Magazines

*AI Expert.* Miller Freeman Publications, San Francisco, CA.

*Byte.* Byte Publications, Peterborough, NH.

*Compute!* Compute! Publications, Greensboro, NC.

*Computer Graphics World.* PennWell Pub. Co., Westford, MA.

*Computerworld.* CW Publishing Inc., Framingham, MA.

*Datamation.* R. R. Bowker, New Providence, NJ.

*Lotus.* Lotus Publishing, Cambridge, MA.

*MacUser.* Ziff-Davis Publishing Co., Boston, MA.

*MacWorld, the Macintosh Magazine.* PCW Communications Inc., San Francisco, CA.

*PC Magazine.* Ziff-Davis Publishing Co., New York, NY.

*PC World, The Magazine of PC Products and Solutions.* PCW Communications Inc., San Francisco, CA.

# Index